图书在版编目（CIP）数据

食品店 /（意）希拉波利尼编；鄢格译. -- 沈阳：
辽宁科学技术出版社，2013.6
ISBN 978-7-5381-7969-9

Ⅰ. ①食… Ⅱ. ①希… ②鄢… Ⅲ. ①食品－商业建筑－室内装饰设计－作品集－世界 Ⅳ. ①TU247

中国版本图书馆CIP数据核字(2013)第054261号

出版发行：辽宁科学技术出版社
　　　　　（地址：沈阳市和平区十一纬路29号　邮编：110003）
印 刷 者：利丰雅高印刷（深圳）有限公司
经 销 者：各地新华书店
幅面尺寸：215mm×285mm
印　　张：14
插　　页：4
字　　数：50千字
印　　数：1～1500
出版时间：2013年 6 月第 1 版
印刷时间：2013年 6 月第 1 次印刷
责任编辑：陈慈良　于峰飞
封面设计：杨春玲
版式设计：杨春玲
责任校对：周　文
书　　号：ISBN 978-7-5381-7969-9
定　　价：198.00元

联系电话：024-23284360
邮购热线：024-23284502
E-mail: lnkjc@126.com
http://www.lnkj.com.cn
本书网址：www.lnkj.cn/uri.sh/7969

FOOD SHOP INTERIOR

（意）西尔维娅·希拉波利尼／编　鄢格／译

食品店

辽宁科学技术出版社

Contents 目录

006 **Chapter One: Overall Design**
第一章：总体设计

014 **Chapter Two: Exterior Design**
第二章：外部设计

020 **Chapter Three: Space Design**
第三章：空间设计

030 Bubble Tease
珍珠奶茶店

034 First Café
第一咖啡

040 La Maison Des Maitres Chocolatiers
大师巧克力之家

046 Sprinkles Ice Cream
Sprinkles 冰淇淋

050 Polka Gelato
波尔卡冰淇淋

054 Sweet Chill
甜冰蛋糕店

058 Café Bourgeois
资产阶级咖啡店

062 Olo Yogurt Studio
Olo 酸奶工作室

068 Melt Me
融化我巧克力和冰淇淋店

072 Pusateri's
普萨特里食品店

076 Raoul's Hammersmith Grove
铁匠林拉乌尔熟食店

080 Pastry Store "Martesana"
糕点店 "Martesana"

086 TSUKIAGE-AN
月扬庵

090 Blè Food Hall
布莱食品店

096 UHA Mikakuto
悠哈糖果店

102 Snog Chelsea
斯诺格冰品店切尔西分店

108 Nana's Green Tea ARIO Kurashiki
七叶和茶仓敷店

114 Nana's Green Tea Uehonmachi Yufura
七叶和茶上本町店

120 Nana's Green Tea Sendai Parco
七叶和茶仙台店

126	*Chapter Four: Food Service Counters* 第四章：食品服务柜台	184	Café Chocolat 诱惑巧克力吧
		190	Nascha's 娜莎食品店
134	Oliver Brown 奥利弗·布朗食品店	196	Candy Room 糖果屋
140	Jewels Artisan Chocolate 珠宝手工巧克力店	202	The Candy Stop Coyoacan 科约阿坎糖果店
144	Little Bean Blue Little Bean Blue 咖啡店		
148	Coffee Hit Hit 咖啡	206	*Chapter Six: Design Guidelines* 第六章：设计规范
154	Cafenatics Cafenatics 咖啡店		
160	*Chapter Five: Food Display* 第五章：食品展示	216	*Chapter Seven: Fixtures, Fittings and Equipments* 第七章：家具、装置及设备
166	William Curley 威廉姆·科利点心店		
172	Bea's of Bloomsbury Bea's of Bloomsbury 蛋糕店	224	*Index* 索引
178	Margaret River Chocolate /Margaret River Providore 玛格利特河巧克力店 / 玛格利特河供应店		

Chapter 1: Overall Design
第一章：总体设计

General Requirements
总体要求

008 Appropriate for Use
合理使用

008 Adequate Space
足够空间

Design and Layout
设计和布局

010 Flow of Food Through a Food Shop
食品生产及消费流程

010 Cleaning, Sanitizing and Maintenance
清洁、卫生和维护

010 Food Preparation Areas
食品制作区

010 Dining Areas
就餐区

Signage Design
标牌设计

011 Customer Service or Directional Signs
顾客服务或指示标牌

012 Signs for Branding
品牌推广标牌

012 Educational Signs
信息标牌

013 Sale Sign
促销标牌

013 New Items
新品推广标牌

013 Monthly Themes
每月主题标牌

CHAPTER ONE OVERALL DESIGN 第一章 总体设计

GENERAL REQUIREMENTS

Appropriate for Use
Food shops design and layout must be well planned taking into consideration several important elements to ensure an effective and acceptable operation. Food shops design principles must accommodate safe flow of product and waste to minimise risks of food and equipment contamination. Separating particular processes in the food shops must be considered including:
a. raw and cooked foods
b. hand washing facilities
c. wash areas
d. storage facilities
e. waste disposal areas
f. toilet facilities

Adequate Space
Proper planning of food shops will effectively designate adequate space and areas for food activities and storage of equipment. Storage areas must be constructive of materials which are durable and easily cleaned in line with requirements for floors, walls and ceilings.
Adequate space must be provided for:
a. food delivery access
b. dry goods storage – sufficient shelving space, pantry area and food

总体要求

合理使用
食品店的设计和布局需要经过仔细地规划并且应该考虑多方面的相关因素，以确保店内能够获得高效的运营。同时，食品店的设计原则还必须包括安全的食品制作流程以及有效的垃圾处理方式等，以避免食品和设备受到污染，需要考虑的因素如下：
1. 食材及食品
2. 洗手设施
3. 舆洗区
4. 存储设施
5. 垃圾处理区域
6. 卫生间设施

足够空间
食品店的内部合理的规划可以确保获取足够的空间用于食品生产与设备存放。此外，存储空间、地面、墙面以及天花板需要采用耐用的和易清洗的材质打造。
必须确保足够的空间用于如下活动：
a. 食品传送通道
b. 干货存储区：

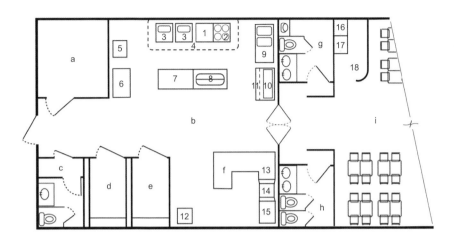

Sample Layout of a Food Shop:
1. Grill
2. Range & oven
3. Friers
4. Canopy hood
5. Handwashing sink
6. Reach-in refrigerator
7. Work table
8. Pan rack
9. Vegetable sink
10. Salad table
11. Under counter Refrigerator
12. Utility sink
13. Pre-wash
14. Dishwasher
15. Clean dishes
16. Coffee maker
17. Hand sink
18. Wait station
a. Storage
b. Kitchen area
c. Staff room
d. Freezer
e. Walk-in
f. Soiled dishes
g. Male
h. Female
i. Dining area

食品店
标准布局：
1. 烤架
2. 炉灶和烤箱
3. 油炸器
4. 排风罩
5. 洗手水槽
6. 冷库
7. 工作台
8. 平底锅支架
9. 蔬菜水槽
10. 沙拉吧
11. 台下冰箱
12. 公用水槽
13. 待洗餐具
14. 洗碗机
15. 干净餐具
16. 咖啡机
17. 洗手槽
18. 等候区
a. 存储间
b. 厨房
c. 员工区
d. 冰箱
e. 冷库
f. 脏餐具
g. 男卫生间
h. 女卫生间
i. 用餐区

grade containers for anticipated stock levels

c. hot and cold food storage – adequate refrigeration, freezer and bainmarie (hotbox) food storage including display areas, food preparation areas and expected deliveries

d. cleaning chemicals and equipment storage – separate lockers, cupboards, cabinets or designated storage areas

e. waste management – sufficient and separate waste containers for all anticipated waste including cardboard, glass, general waste, and waste oil storage; waste bins must be impervious, and designed to be easily cleaned to prevent the attraction of pests. Waste bins washing/cleaning area or room that complies with the requirements of the trade waste and the Environmental Protection Act must be provided when required for cleaning waste bins

f. personal belongings storage – separate lockers, cupboards, cabinets or designated storage areas

g. food contact utensils storage – adequate storage containers that can be easily cleaned as well as preventing contamination

h. equipment storage – sufficient floor, cupboard or shelving space for all cooking and food preparation equipment to be stored

i. food packaging material storage – adequate storage located off the floor and protected from contamination

j. office and business equipment (used to run the business) – must be separate from the food storage and preparation areas to prevent contamination

货架区、茶水间、食品级容器、
（用于食品分级存储）存储区

c. 冷食及热食存储区：
制冷设备与冷冻设备、加热设备存储区

d. 清洁用品及设备存储区：
独立的储物柜、橱柜等摆放区

e. 垃圾管理区：
分类垃圾存储容器区域；
垃圾箱须防渗漏；
垃圾箱须易清洁，防止招引害虫；
垃圾箱清洗区须遵循垃圾分类标准等

f. 个人物品存放区：
独立的储物柜、可用于存放物品的其他容器或空间

g. 食品盛放器皿存放：
足够的存放空间、空间易于清洁并防止污染

h. 设备存放区：
足够的区域用于存放烹饪及食品制作设备

i. 食品包装材料存放：
足够远离地面的存放空间，防止污染

j. 办公区及办公设备区：
与食品存放和制作区分离，防止污染

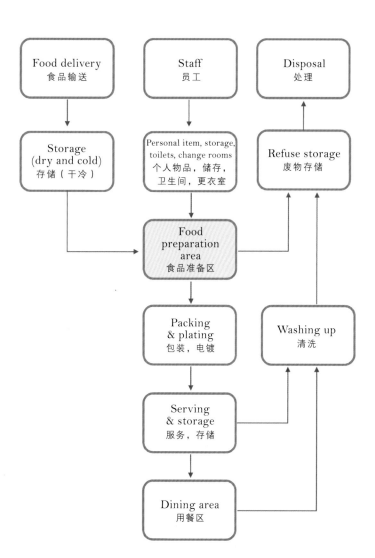

Left: Flow of food through a food shop
左图：食品生产及消费流程

DESIGN AND LAYOUT

Flow of Food Through a Food Shop
The correct design and layout can help streamline work practices, reduce cleaning and maintenance and prevent cross contamination.

To prevent food contamination, shops must be designed so that the flow of food is in one direction from receipt to storage, preparation, packaging and serving and finally to disposal.

Cleaning, Sanitizing and Maintenance
Layout and design of the shops must provide access for cleaning, sanitizing and maintenance.

Food Preparation Areas
Adequate space must be provided for all food related activities. Exits must be in accordance with the building code of the local government.

If the shop is an existing building, one may need to check with a building certifier to see if the exits comply.

Dining Areas
Adequate space must be provided for patrons and staff to access dining and serving areas.

设计和布局

食品生产及消费流程
合理的设计和布局可以帮助优化工作流程，减少清洁及维护工作。

为了防止食品受到污染，店内人员必须严格遵循食材的存放与制作、食品包装、供应以至废物处理等流程方式。

清洁、卫生和维护
食品店内必须遵循清洁、卫生及维护等相关标准。

食品制作区
店内的空间设计必须为与食品相关的所有活动提供足够的空间。同时，出口设计也需要严格遵循当地的建筑标准。

如果食品店选址在既有建筑内，应向相关人士确认，出口设计是否符合要求。

就餐区
店内必须提供足够的空间，能够用于顾客和工作人员的日常就餐。

Above from left: Sign/signage design
以上从左至右：食品店标志 / 标牌设计

SIGNAGE DESIGN

Signage is one sales promotion effort that is guaranteed to get results. University and wholesale group reports show that better signs increase sales and profits. Surprisingly, many retailers don't devote enough time and attention to this marketing component.

A professional sign is vital. There's nothing worse than a hand-written, crumpled, faded sign hanging lopsided. It ruins the professional image. Be consistent with the feature lines, price size, colour and fonts so the customer's eyes can easily scan the sign. Allow for plenty of white space, keep the font simple, be sure the text is balanced and proportional, and consider using bullets. Highlight words in bold or by using a different colour. Avoid using all capital letters, which makes it difficult to read.

When there is a lot of information on the sign, design it to read from left to right by having the text start on the left side rather than in the centre-e.g., a 3-line description is harder to read if it's justified to the center. Also, be as specific as possible.

Customer Service or Directional Signs

These can be gondola signs, inline shelving signs or on-shelving signs; and they're used as directionals. You can have the best customer

标牌设计

标牌设计可以作为营销推广的手段之一。来自大学及其他团体的一些报告曾指出，好的标牌设计可以提高销售额和利润收入。然而，许多经营者却忽略了这一点。

一个专业化的标牌对食品店的经营起着至关重要的作用。对于一家食品店来说，没有比一块字迹潦草、悬挂不当的标牌更能破坏其自身形象了。标牌设计应该保持特征线、颜色及字体的一致性，以便于顾客准确地阅读。同时，还应该保留一定的空白，保持字体简约，确保文字内容平衡并且适当使用着重符号；突出关键词或使用不同的颜色，避免使用大写字母。

如果标牌要展示很多信息，确保文字从左侧起排，而不是从中间位置开始。例如，占据三行的信息内容如果从中心位置起排，那么阅读起来就会非常困难。同时，内容表现信息要具体。

顾客服务或指示标牌

货架标牌被视作指示标识的一种。一些食品店可以提供良好的顾客服务，但是，当店内的服务人

Above: Website design of a shop
上图：食品店网站设计

service going, but what happens when you have 4 employees with customers and a fifth customer walks in? You always want to acknowledge a customer when they walk in the door. A directional sign can serve as a key part of your customer service by promptly steering the customer to a category or aisle until an employee can reach them. Easy to read aisle markers is a great start, but you need more. You can have an aisle marker that says "VITAMINS", but think of how the customer feels once she gets to your vitamin aisle. It looks a mile long to the customer! You want her to easily find what she wants.

Signs for Branding
Almost like a subliminal message, your logo, tag lines and web site address should be consistent on all of your signs. (Image Above)

Educational Signs
Signs provide an opportunity to share information: announce an upcoming event in your store, discuss a hot issue in the news, or entice customers to frequent your web site. You might want to print a recipe from your web site onto a sign that says, "Get this recipe and more on our web site." You can do the same thing with the news item (invite them back for more). According to Danny Wells, "7% of customers that shop at natural food stores turn to the Internet for health and wellness information." Why not steer them to your web site?

员的数量无法满足所有顾客需求的时候，指示标识就可以帮助改善这一状况：使用清晰、易读的过道标牌是一个很好的开端，但是仍旧需解决很多的问题。举个例子来说，顾客可以沿着店内标明的"VITAMINS"（维他命）的过道标牌找到自己所要购买的商品类别，但是，那些摆满货品的长长的货架仍然会或多或少地使他们感到困惑。因此，在指示标志的设计上首先需要改进的便是使其清晰易懂，以便顾客能够一目了然、轻松地找到自己所需的产品。

品牌推广标牌
犹如潜在信息一般，标牌上的品牌标识、标签、网址等信息要完全保持一致。

信息标牌
标牌可以帮助共享各种信息，如店内将要举办的活动、对于热点问题的讨论以及吸引顾客访问店铺网站。可以从网站上复制一条菜谱贴到标牌上，然后写上"赠送一条菜谱，在本网站上可获取更多"的字样。当然也可以利用这种方式推销新产品。丹尼·韦尔斯曾指出，7%去生态食品店购物的顾客都会上网查询关于保健方面的信息。那么，可以考虑在店铺网站上提供相关信息，吸引顾客访问。

Above: Graphic design of a shop
上图：店内平面设计

Above: Signage design of a storefront
上图：店面标牌设计

Sale Sign

At a surprisingly low price, you can increase your sales significantly with effective signs. Give the customer a reason to pick up the product. Find a sizzle statement to attract the customer to the sale item-e.g., fresh organic herbs, wheat free, grown locally, fresh pressed, etc. Remember, it's not always about price. Use your signs to guide customers to the sale items: "See the values in Aisle 5." Give strong buying commands on signs, flyers and coupons: "Look at These Features," "See the Values" or "Don't Miss This Opportunity!"

New Items

Use signs to create excitement and educate the customer about new items. You can bring the greatest new item into the store, but if you don't have good signage or give out some information, it's the best kept secret! You've spent the money to bring the product into your store, so create a little excitement! You might want to display new items on end-caps and see how they do before you put them into the inline shelving.

Monthly Themes

Develop a themed calendar for the year, then create an end-cap display with 12 different signs that you can easily rotate each month. It's reusable, saves you money and can bring a fresh look to your store throughout the year.

促销标牌

运用促销标牌可以有效地提高食品店的食品销量。为顾客找到一个购买此食品的充分理由，并且采用一些极具说服力的词句吸引顾客前来光顾，比如一些新鲜有机食物、无麸质、本地种植、新鲜压榨的水果等。由于价钱并不是决定购买的唯一条件，有趣的标牌内容能强化顾客的购买欲，并且标牌的内容可以多种多样，如"去第五排货架寻找价值"、"看看这些特色品"、"请勿错失良机"等。

新品推广标牌

运用标牌告知顾客店内的新品是食品店比较普遍采取的方式，并制造一些特色来提升食品店在顾客心中的位置。如果只是单纯地把产品引进而来，但不将信息发布出去，会造成信息闭塞，顾客不能够及时知情，对销售来说毫无意义。因此，可以将新品放在显而易见的地方进行出售，观察销售结果，然后再上架售卖。

每月主题标牌

制作一个由12个不同标牌组成的年度特色主题日历，保持每月更换，既可以重复地使用，又能够节约预算。一年中的每个月，食品店内都会呈现出新的面貌，增加消费者的新鲜感。

Chapter 2: Exterior Design

第二章：外部设计

Site Design
店铺选址

Storefront Design
店面设计

016 Colour
色彩

017 Material
材质

017 Lighting
灯光

017 Colour Contrast
对比色

017 Design Contrast
对比风格

018 Storefront as a Design Element
店面作为一种设计元素

018 Symmetry Versus Asymmetry
对称性与非对称性

018 Open Façades
开放式立面

018 Transparency
透明度

018 Outside / Inside
室内外融合

018 Canopies
顶棚

019 Banners, Flags and Signs
广告牌和标识

019 Street Signage
街道标牌

019 Branding on Storefront
品牌标识

019 Hoardings
招贴板

019 Fragrance and Smells
香味

Above: Corner is often preferred in the food shop site design
上图：在食品店选址中，街角通常是首选

SITE DESIGN

Sites for food shops must be chosen that are free from conditions that might interfere with their sanitary operation, including:
1. No land use conflicts or potential conflicts with adjacent sites;
2. Set reasonably apart from waste disposal facilities, incompatible processing facilities, and any offensive trades. Generally a minimum set back of 30m^2 is recommended from potential sources of contamination. However, a greater or lesser distance could be accepted based on specific site conditions.
Surrounding facilities must not contaminate food. Conditions which might lead to contamination include excessive dust, foul odors, smoke, pest infestations, airborne microbial and chemical contaminants, and other similar conditions.
Checklist for location selection as below:
- Explore pedestrian movement patterns
- Vehicular traffic patterns
- Estimate times of peak activity
- Demographics of potential customers
- Image of the area – high end or budget
- Survey of other stores in the area
- Estimate of customers if within another store or complex

STOREFRONT DESIGN
Colour

Using colour is a very simple and powerful tool in retail. Vibrant colours

店铺选址

食品店的选址必须满足卫生安全的要求，具体包括以下两个要点：
1. 不能与周围的环境存在冲突或潜在冲突；
2. 所在的位置要远离垃圾处理和加工场所。在一般的情况下，具体的要求是要与潜在的污染源最小距离应该达到30米。但是也可以根据具体情况，决定距离的多少。
同时，要确保选址周围的环境一定不能对食品店的产品造成污染，主要是远离灰尘、烟尘、昆虫侵害等污染源。
其他需考虑元素如下：
- 行人流动模式
- 交通流动模式
- 预估人流高峰次数
- 调查潜在顾客类型
- 研究区域发展现状（高端或普通）
- 观察同区域内同类店铺状况
- 预估同类店铺顾客数量

店面设计
色彩

色彩选择应注重简约有力。红色和橙色等较为活泼

Above: White and black are often used as contract colours in the storefront design
上图：白色和黑色是店面设计中经常用到的对比颜色

Above: Ourdoor lighting plays an important role in the storefront design
上图：室外灯光在店面设计中起着非常重要的作用

like red and orange are instant attention grabbers. Colours like green and blue project elegance and sophistication.

Material
The use of different materials can send "messages" to customers. (eg., using wood exudes elegance; metals imply modernity and sophistication.)

Lighting
Lighting creates an extremely forceful statement. Although lighting can be used during the day also, it is most successful in the evening and night. Good lighting design can not only provide a means to create focus, but also to create the perception of safety and comfort.

Colour Contrast
Using contrasting colours can be very useful. This can be used for store window displays, exterior signage and interior. Typical examples will be the use of white and black, red and black and white on a dark background etc.

Design Contrast
This technique may be useful in historical districts or on busy retail streets. When there are a large number of stores and you need to stand out, this hot button can prove very handy. Design contrast can be achieved through architectural details and use of materials.

的颜色能在短时间内吸引顾客的眼球，而绿色和蓝色则突出优雅和高贵。

材质
不同的材质可以向顾客传达出不同的信息，如木材散发典雅和魅力，金属凸现代和精致。

灯光
灯光往往能够营造一种强有力的氛围。尽管灯管在白天也可以运用，但其在傍晚和夜间能打造更好的效果。好的灯光设计不仅可以引人注目，更能营造安全感和舒适性。

对比色
采用对比色往往能够达到更好的效果，可以用到展示橱窗、室外标牌的设计上。最为典型的例子即为黑、白色调的同时运用，红色、黑色和白色共同出现在一个背景中。

对比风格
对比风格往往运用在位于古老历史区或繁忙大街的店面设计中。如若想在周围的环境中脱颖而出，则可使用对比风格，通过细节和材质的运用达到预想的效果。

Above: Architectural shape of the storefront design
上图：店面设计中的建筑元素

Above: Open façade of the storefront design
上图：开放式的店面设计

Storefront as a Design Element
The storefront can actually be designed as an architectural shape. This will help to differentiate from other stores and make customers remember.

Symmetry Versus Asymmetry
Symmetry creates a sense of peace and can be used in providing a "pause" on a busy street front. Moreover, asymmetry can provide excitement.

Open Façades
This can be seen as an extension of transparency. By having no "storefront" the retail space is easier to explore – customers need not open a door to get in.

Transparency
This is one of the most powerful tools available in retail design. By using large expanses of glass, customers can see most or all of store before they enter it. Glass creates an extremely inviting retail experience.

Outside/Inside
This helps to integrate the extension of store with the interior. Examples include products on sale on the sidewalk outside store and sidewalk cafés.

Canopies
Canopies can be used successfully to attract attention and create an interesting storefront, and also be lighted at night to create a statement.

店面作为一种设计元素
店面可以设计成一种建筑造型，这样就可以与众不同，给顾客留下更深的印象。

对称性与非对称性
对称性能够营造一种平静的感觉，在繁忙的街道上能够让人平和，而非对称性则会带来跳跃感。

开放式立面
开放式的立面可以被视作是"透明度"的延伸，营造出"无店面"的感觉，这对顾客来说更加方便。

透明度
透明度是商店设计最为有效的要素。通过大幅玻璃的运用，顾客在进店之前就可以看见大部分商品。玻璃能够营造温馨的购物体验。

室内外融合
室内外空间的融合对于店面设计格外重要。举个例子，可以将食品拿到店外售卖。

顶棚
顶棚运用可以成功吸引行人的目光，趣味性十足，同时夜晚在灯光的照耀下更具特色。

Above: Transparency of the storefront design
上图：透明性的店面设计

Banners, Flags and Signs

These are again time-tested mechanisms to attract attention, especially on retail streets. Combined with some other hot buttons like colour, these are successful and easy to use. Their greatest advantage is that they project into the field of view and are hence extremely popular on streets.

Street Signage

Another mechanism to attract attention is to install freestanding signage on the street right outside store. The simplest is a sign that can be brought inside at the end of the day. More sophisticated examples would be signposts erected on the sidewalk.

Branding on Storefront

Branding can be achieved with affixing logo onto the storefront. This combined with transparency can create a powerful retail experience.

Hoardings

This hot button can be extremely effective for corner locations.

Fragrance and Smells

This is not a very frequently used tool. Due to bylaws and code restrictions, this might be used only in limited exterior locations. Food shops can use this by exhausting onto the street. So even if pedestrians fail to see the store, they will smell it!

广告牌和标识

这些元素结合丰富的色彩运用，则更能打造出非同一般的特色，尤其是当被运用在繁荣的商业大街上。它最大的优势还在于能够足以吸引眼球并大受欢迎。

街道标牌

在店外的街道处树立独立的标牌同样可以吸引路人的注目。最简洁的方式是制作那种可以在非营业时间拿到店内的标牌，较为复杂的做法则是人行道上的路标牌。

品牌标识

品牌标识可以运用到店面标牌上，与透明度结合可以打造强烈的购物体验。

招贴板

招贴板多用于处于拐角位置的店铺。

香味

这一元素并不常用。根据相关法规，只能在有限的室外空间中运用。食品店的运用相对多一些，尽管行人可能看不到店铺，但通过店铺由内而外散发出的香味也可以为此驻足停步。

Chapter 3: Space Design
第三章：空间设计

Interior Design
室内设计

022 Interior Layout
室内布局

023 Holistic Design
整体设计

024 Colours
色彩

024 Materials
材质

025 Furniture
家具

Lighting
照明

027 Entrance
入口

027 Market Place
主卖场

027 Main Aisle
主过道

028 Aisle with a Range
陈列食品的过道区

028 Rear Wall
后墙

028 Cash Desk
收银台

028 Bread and Pastries
面包和糕饼区

029 Meat, Sausage and Cheese
肉类、香肠和奶酪区

029 Deep Freezers
冰柜区

029 Wine
酒品区

CHAPTER THREE SPACE DESIGN 第三章 空间设计

INTERIOR DESIGN

Interior Layout

Being an architect primarily means learning the activities and attitudes of those who will be the final users of the space. You may therefore need to be in the shoes of a bartender, confectioners, ice-cream makers, dealers, waiters and cooks. It's vital to look through, to listen, feel, touch and taste. Designing food shops requires different skills compared with any other space definition. First of all you have to consider carefully the complexity of a public space that is always heavily linked to safety and accessibility requirements. As for the analysis of the functions, it's mandatory to elaborate a correct layout that has to withstand situations of overloading, channeling the flow of people, food or instruments without overlapping each others. The definition of interior layout is more and more including the idea to disclose the "backstage" where food is prepared with the intention to show the incredible attraction that the elaboration of the food has The visual continuity between the place can satisfy customers' curiosity. So it's possible the backstage is more visible and closed to the sales space, both geographical and physical, as a guarantee of quality.

Working spaces are essential in the layout definition as the technical implants project. An immediate integration of them in the project enables the development and the resolution of all the potential issues related to these complicated instruments both in work rooms and sales areas.

室内设计

室内布局

作为一名设计师，必须了解食品店空间最终使用者的活动和态度，可以把自己想象成一名调酒师、糖果店店员、冰激凌大师、房产经纪人、餐厅服务员或大厨，要不断地去观察、聆听、碰触和感受。

同其他类别的室内空间相比，商店设计需要更多的技巧性，而食品店设计的要求则更高。首先，食品店作为公共空间，必须考虑其安全性和实用性两方面需求。其次，在功能性方面，必须打造一个合理的空间布局用于不同种类商品的分区和摆放，以避免出现人流拥挤、食品及设备罗列的现象。

在食品店的空间布局中应该体现出其"后台"，即食品的制作空间，以便于向顾客展示食品的制作过程，吸引消费者的购买欲。食品的制作空间和售卖空间在视觉上的连续性可以在很大程度上满足顾客的好奇心，但是同时也要注意两个空间的间隔性，以确保食品的卫生安全。

工作区在食品店中的位置同样重要，这一区域的设置能解决所有与工作设备和销售区域相关的问题。

Left: Sign and package designs are important to the holistic design
左图：标识与包装设计是整体性设计中的重要部分

Holistic Design

A successful project is not simply the easy sum of its contents. A space is involving when it's appreciated as a whole, not like the addition of several elements each others unlinked. The part of the project that is not recognisable as specific object but let the space to be appreciated like a whole identity, is what designers define the concept. Find a strong idea, a content, a message to be sent to the people, is the first step to an holistic approach to the design. The concept is the overall "rule" to be applied to all aspects of the project: space definition, relation between different parts, products exhibition, holders, packaging, sign boards, furniture design. Everything has to follow a file rouge conducting to the whole. In this way, you get a 360° holistic design that takes in account all the aspects, even those that apparently seem not to be connected each other. Architecture when is experiencing the world of food, has to examine many different aspects, as a huge math equation. You have to evaluate the needs of people that work, buy and sell in the same space, integrating them with disciplines similar to architecture as illuminating engineering, plant design and physics. Interaction of above described system with external environment is important both from a naturalistic point of view and for what concerns the mutual influence with human life. The scope of this method is the wellness of the final users. Aesthetic, in this case, is not just a dress to be applied in a second time, but it's the outside aspect of the concept that reveals itself through images and that enables to perceive the project as the whole.

整体设计

一个成功的设计并不只是各种视觉设计元素的简单叠加，一个足够吸引众人的空间应该是看这个空间的整体效果。通常来讲，设计的宗旨是即便空间中的每一个部分都不足以引人注目，但是也必须具备整体统一的特质。整体设计概念的起点即为传递一个强烈的信念、一段丰富的内容或者是一条清晰的信息。其种的设计规则可以运用到各个方面，比如空间定位、不同元素之间的关联、产品展示、包装材料、标识牌及家具设计等，其中每一部分的元素都必须遵从这个"整体"原则，只有按照这个规律，才能够打造出一个全方位的有效空间。当建筑空间成为一个售卖食品的"世界"，则设计就应该考虑包括食品以及其他众多方面的客观元素，比如店内的工作人员、顾客等的需求。并且还需要将所有这些元素与建筑相关联的其他学科融合在一起，包括工程设计、装置设计等。同时，从自然科学以及同人类的相互影响角度出发，以上这些理念还应该同外界的环境联系在一起，比如地理位置、周围的交通等。无论如何，上述的理念须始终从空间最终使用者的角度出发，确保满足他们的意向和需求。除此之外，美感也是影响空间使用的另一元素，它不仅仅扮演着装扮者的角色，更是体现整体空间形象的要素。

Above from left: Interior colours design
上方两图：室内色彩

Colours

The chromatic study and the application of the basic colours theory are the base for realising comfortable environments. Within a space, colour may differently influence the perception of place. Interior designer is expected to fully evaluate the final use of the space in order to choose appropriate colours.

There's a tight relation between food and colours where it is bought or consumed. Space shades are prominent and may generate various sensations depending on the tonality, for example, colours of brown and beige are often used for shops selling candies; on the contrary, orange, red or green may better fit for presenting salt food. Generally speaking, black, red and white mentally increase food desirability. Neutral and hot colours are more suitable for selling food.

Light is basic for empowering the natural colours of the products. Whenever it's possible, it's advisable to use natural light that has anyway to be complemented with the artificial one. This last one should support in disclosing the products as the real main characters of the space, creating pleasant and enjoyable atmosphere.

Materials

Materials in a food shop have a strategic role and have to be picked with the biggest care. As you are projecting a public space, materials

色彩

色彩研究以及基本色彩应用理论为打造一个舒适的空间环境奠定了基础。在一个空间内，色彩能够在很大程度上影响人们对此环境的认知感。因此，设计师应该根据空间的最终使用功能来选择色彩基调和相关的搭配。

出售中的食品和空间色彩有着密切的联系。由于色调在空间设计中至关重要，不同的色调可以产生不同的环境氛围：比如糖果店中多运用褐色和米色，而红色、黄色和绿色则多数是用在高盐食品店中。一般说来，黑色、红色和白色可以提高顾客对于食品的购买欲。此外，中性或者较为浓重的色彩更为适合。

灯光是凸显食品天然色彩的基本元素。在条件允许的情况下，尽量选择利用天然光线和人造光线相结合的形式来烘托食品的自然效果，使其成为空间内的主要特色，从而营造一个宜人的环境，使消费者对食品的新鲜程度有一定了解。

材质

在食品空间设计中，选择恰当的材质起着至关重要的作用，因此需要根据食品店的特征和需求做精心

Above: White Calacutta marble slabs are set behind the meat counter
上图：白色的卡拉卡塔大理石板被安放在肉类柜台的后方

Above: White corian is used to express the creamy quality of ice cream
上图：白色可丽耐被用来表现冰淇淋的奶油特征

should measure up with resistance and cleaning requirements, mainly when you have to cope with scrap and dirty resulting from food preparation. Materials have to be selected accordingly to the concept for obtaining the visual impact, colours and the texture desired.

Visual neutrality is a factor to be considered in choosing materials for showing the foods. For example proposing the food on a veined marble could create visual confusion and draw the attention from the object of sales.

Glass may be used for different solutions: its natural transparency lets see through and at the same time protects and keeps divided from the rest. It's commonly used for display cabinets, brackets, furniture or just like a baffle for separating the kitchen and production space from the shop.

Preferring the absence more than the presence of materials does not mean to meet a kind of radical minimalism, but it's the wellness to emphasize the hierarchy of the elements to give the products the importance and the visibility they deserve.

Furniture

Furniture in a shop have clearly a core relation with the activity and user's expectations. Therefore, it's critical to fully understand the

选择。例如，在公共销售空间内，材料的选择应该多侧重在质地的耐用性和易清洁性上。尤其在对食品备料中产生的碎屑和脏物进行处理时。此外，材质的选择也应充分考虑视觉影响、色彩及质感等。

在材质选择上，营造中性氛围至关重要，以便于突出食品。例如，将食品陈列在纹理丰富的大理石上，引发的视觉混乱容易将顾客的注意力转移到材质本身。

玻璃是一种常用的材质，它的天然的通透感既不影响视觉（又对视觉效果有着积极的影响），又能起到对食品的保护作用。因此通常用在展示柜、货架和家具上。在食品店内，将厨房和备料区与商店分隔开来。

尽量减少材质的展示性并不意味着要完全遵循极简主义的设计风格，而是要突出并强调空间元素的层次性，比如高度、深度等，从而赋予食品更多的可见性和吸引力。

家具

在食品店空间中，家具同店内的宣传活动和使用者的意愿联系十分密切。因此，设计者应该充分考虑

Above: Lighting design of the entrance
上图：入口处灯光设计

Above: Lighting design of the main aisle
上图：主过道灯光设计

needs that furniture is expected to fill and consequently evaluate if the product is already available in market. If not, after due budget consideration, it's almost interesting to be in the position to think of a totally customized solution with the idea in mind to propose a brand new way of utilization and exhibition.

In this contest details are essentials for proposing a real functional furniture. Learning movements and operations daily done in the space and fitting the furniture into them enable to reduce utilization time and facilitate sales process, ensuring a high level service.

The demand to expose products sometimes comes up against the necessity to preserve the food refrigerated or warmed or to cover it from polluting agents. Technology is indispensable for fixing above requirements. (Author: Silvia Cirabolini, StudioAlbertoRe, Bibliography reference: Piersanti, R.Rava, "*Architettura e gusto*", Electa, 2007)

Lighting

The setting of lighting scenes for food is one of the most demanding tasks. Particularly in the case of fresh products, quality lighting is of the highest importance. Good light underlines the quality of food and preserves it. In the case of every product-specific lighting solution, lighting which is gentle to the product is therefore at the forefront.

家具对空间所能产生的作用和是否能够满足店主和顾客的需求，在这个前提下，设计者才能明确地在市场上寻找到合适的家具类型。如果遇到没有符合所需的类型，那么在预算允许的情况下，可以考虑有计划的定制。

打造一件极具实用性的家具应充分考虑各种细节元素，了解店内日常活动和经营状况，使其完全满足空间要求，减少使用时间，优化销售流程，从而提供高品质服务。

在展示食品的同时也会遇到如下问题，如何保存冷藏或加热食品，如何避免食品受到污染。为解决如上问题，运用一定的技术则显得尤为重要。（本文由 Silvia Cirabolini/StudioAlbertoRe 工作室提供，参考资料：《建筑的味道》）

照明

利用灯光来营造背景，在食品店设计中是一项重要而又棘手的工作。对于烘托新鲜食品来说，高质量的照明则显得尤为重要。良好的照明既能突出食品的品质，又能保护其不受损害。不同食品对于照明的效果有着不同的要求，而柔和的灯光条件则往往被置于首位。

Above: Lighting design of the rear wall
上图：后墙灯光设计

Above: Lighting design of the cashdesk
上图：收银台灯光设计

Modern lighting concepts set an attractive and sales-promoting scene for the goods. They create a pleasant shopping atmosphere and sharpen the profile of the dealer's brand. This requires a high degree of professionalism and sophisticated know-how.

Entrance

Involved here is the attractive distant effect with recognition value and the inviting entrance area: the entrance characterises the shop to the outside. Its effective function goes far beyond the functional building access. Here independent luminaires of a decorative nature which have a striking and high-quality appearance prove successful.

Market Place

The lighting for the fresh products which are shown in all colour facets should have an appetite-whetting effect and be gentle on the merchandise.

Main Aisle

Along a main axis the light serves in particular for orientation, shows the direction and leads people into the adjoining aisles and areas. The designer systematically looks after this important main access axis with the help of an appropriate gondola head lighting system, arranged along the aisles and at the same time setting the scene for goods in an attractive way.

现代的照明理念旨在营造迷人、舒适的存在环境。店内主要是突出食品的品质，从而为顾客带来信任感和愉悦的购物体验。这就要求具备高水准的专业素质。

入口

入口，即门面，是一家商店的外部形象，不仅仅具备建筑与外界连通的功能性，迷人的远景效果以及独特的品质，更易吸引路人。极具装饰特色的独立照明设备更容易营造高品质的外观，成为入口处的成功设计。

主卖场

新鲜食品的展示能够突出店内的整体性和色彩，照明光线应柔和并能烘托其品质，让顾客产生购买的欲望。

主过道

主过道区的照明主要起到引导和指示作用，将顾客引领到周围近的售卖区。通常情况下，设计师会首先关注主过道的客观条件来确定选用适合的货架端头照明系统，并且沿着过道布置，为食品营造迷人的灯光氛围。

Above: Lighting design of the cakes
上图：糕点区灯光设计

Above: Lighting design of the meat
上图：肉类区灯光设计

Aisle with a Range

Aisles containing ranges are the largest part of the area to be lit. Here the focus is placed on lighting in keeping with the merchandise and on economic efficiency. The accentuated shelf and aisle area lighting is an essential aspect of lighting design. It directs the gaze of the consumers systematically towards the ranges of goods on the shelves.

Rear Wall

An objective is the rear wall illumination, flushed with the ceiling, which appears to cover the whole area and nevertheless appears lively. Here luminaires with asymmetrical reflectors are necessary which, with high efficiency, project the light evenly onto the shelves.

Cash Desk

The cash desk area demands an individualized lighting concept since here is non-glare workplace and lighting, scene setting for goods and orientation aid come close together. In the sales room itself, the cash desk must be visible for the consumers from a great distance.

Bread and Pastries

Bread and pastries are handled with care and they are displayed in a sales-promoting manner. The best way to do so is with a range of appealing warm colors lighting corresponding to the nature of this item of food.

陈列食品的过道区

陈列食品的过道区占据着食品店内照明的绝大部分，该区域的首要任务是确保光线与食品的融合，同时确保最佳经济效益。重点货架与过道区照明构成整体照明的重要部分，可以将顾客有效地引领到货品区内。

后墙

后墙的照明要求要与天花板的造型相呼应，以便涵盖住整体区域。带有不对称反光镜的灯具在这里显得格外重要，它可以确保光线能够均匀地照射到柜台上。

收银台

收银台区要求具备一个独特的照明理念，因为这里不仅是一个无眩光区域，同时也是背景照明和指示照明的聚集地。在售卖区，收银台必须保持一定的可见性，使得顾客从很远处就能望见这里。

面包和糕饼区

面包与糕饼需要经过面点师的精心处理，并以一种美好、诱人的方式陈列。最为有效的方式即为运用暖色调来烘托，与食品的自然品质相互呼应，使消费者能够感觉到食品的温暖与美味。

Above: Lighting design of the deep freezers
上图：冷冻区灯光设计

Above: Lighting design of the wine
上图：酒类区灯光设计

Meat, Sausage and Cheese

Meat as fresh food is also very sensitive. For that reason, the focus is on the least damaging but sales-promoting presentation by means of a combination of efficient lighting equipment with special filters.

Deep Freezers

Deep freezers require homogeneously bright horizontal lighting. The brand images of the goods must be illuminated richly in contrast and without any thermal load. Suspended luminaires place room-forming and decorative accents. They make it easier to find your way and permit visual allocation of the groups of products.

Wine

In this high-quality area what it's all about is the mood. Accentuated lighting increases the attractiveness and the apparent value of the products. Sun filters ensure a warm lighting situation with a minimum of thermal load, in which the consumers like to linger.

肉类、香肠和奶酪区

肉类食品具有很强的敏感性，因此在照明设计上要避免对其造成损害同时确保具备一定的吸引力。可以选择使用带有独特光线过滤装置的节能照明设备组合。

冰柜区

冰柜区需要均匀而明亮的水平照明，突出不同品牌食品之间的对比，同时确保零热负荷。垂悬的灯具装饰性十足，既可以起到指示作用又可以在视觉上将食品分类。

酒品区

这是一个要求高品质的区域，氛围营造至关重要。重点照明可以提升产品的外观吸引力，光线过滤既可以保证温馨的氛围，又能确保最少的热负荷，为顾客营造一个乐于流连的场所。

Bubble Tease
珍珠奶茶店

Designer: GH+A
Location: Mississauga, Canada
Project year: 2010
Photographs: © Philip Castleton,
Philip Castleton Photography Inc., Toronto, ON

设计师：GH+A
项目地点：加拿大，米西索加
完成时间：2010年
摄影师：© 菲利普·卡斯尔顿；菲利普·卡斯尔顿摄影

Bubble Tease serves specialty Asian inspired bubble tea and fresh fruit drinks. The designers re-defined the attitude and brand message of Bubble Tease to guide them in expanding their offering to include fast casual and small plate food items within a destination that would invite customers to socialise and stay a while.

The café design includes various seating options. The large circulation areas at the front portion of the café are designed to accommodate long customer queuing, so customers who are not consuming their beverage or food at the café can wait comfortably. The Bubble Tease bar is the theatre of the café where customers can view all menu items on live feed monitors and see their drinks and food being prepared.

In one area of the café seating, customers can sit at bar height tables surrounded by café "Bubble Tease Geology" wall graphic, which leads a supporting cast of "Personali-tea" flavours on the menu, and includes amusing large scale presence in both the men's and women's washrooms. In another area, customers can sit and lounge in booth type seating with metal chain privacy screens. And the lounge space at the rear of the café is designed to be very flexible so that the area could accommodate a larger gathering as well. The modular furniture in the lounge can be configured a variety of ways simultaneously.

1

珍珠奶茶店专供亚洲奶茶和鲜果饮品。设计师重新定义了该店的品牌形象，引导他们扩展经营范围至快餐和休闲小食，提供顾客驻足社交的空间。

奶茶店的设计包含各种座位的设置。前部的大型流通区域能够容纳顾客的长队，让外卖的顾客能够舒适的等候。珍珠奶茶吧是店内的剧场，顾客可以现场观看所有菜单产品，看到饮品和食物是如何准备的。

就餐区的一角，顾客们可以坐在吧台前，四周的墙壁上绘满了菜单上的饮品风味。男女洗手间的墙壁上还有大型壁纸图案。另一个区域，顾客可以坐在卡座式的座椅上，外面遮挡着金属链条屏风。奶茶店后部的休闲区设计比较灵活，还可以举办较大的聚会。休闲区的组合家具可以任意组合。

1. Lamp design
2. Dining area
3. Service counter
4-5. Dining area

1. 灯具设计
2. 用餐区
3. 服务柜台
4、5. 用餐区

Floor Plan:
1. Entry
2. Seating
3. Retail
4. Ordering
5. Queue
6. Bubble tease bar
7. Kitchen
8. Café booth/lounge
9. Café bar height seating
10. Work station
11. Lounge
12. Storage
13. Back of house

平面图：
1. 入口
2. 就餐区
3. 零售区
4. 点餐区
5. 排队区
6. 珍珠奶茶吧
7. 厨房
8. 咖啡/休息
9. 咖啡吧高脚椅
10. 工作台
11. 休息区
12. 储藏室
13. 后厨

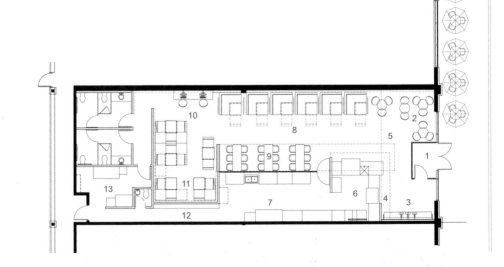

第三章 空间设计_ 033

First Café
第一咖啡

Designer: Pereira Miguel Arquitectos
Location: Lisbon, Portugal
Project area: 170m²
Project year: 2009
Photographs: © Fernando Guerra + Sérgio Guerra
设计师：佩雷拉·米格尔建筑事务所
项目地点：葡萄牙，里斯本
项目面积：170平方米
完成时间：2009年
摄影师： © 费尔南多·格拉+塞尔吉奥·格拉

First Café is a new commercial project located at the old Lisbon airport in Portugal, occupying a total area of 170 square metres. It is located in the centre of the luggage atrium room, serving coffees and toasts to hungry and tired travellers at the airport.

Previously, the space had nothing to offer. People would wait up to one hour for their bags without anything to spend their euros. There were only metal columns pullulating the room in a cellar type of ambience. It was the typical "in – transit" space, with nothing different from other similar airports.

The biggest challenge to this interior project was to create something that could use columns in its favour. At the same time, it should be an open store without closing and even more the existing space. The designers from Pereira Miguel Arquitectos, Lda. have designed a concept that uses columns as an advantage, to store materials and shelves and to provide multiple surfaces to be decorated. All the decoration is based on the idea of having a "sense of place", something related to Portuguese cultural symbols, yet in a modern reinvention. The great paradox is that: a sense of place in a transit space.

To reinforce the idea of the columns, the designers of Pereira Miguel Arquitectos have designed small selling trolleys that go around the airport selling goods and coffees. They are reduced columns in a mimic operation. It should look like all columns could walk as well, leaving the impression that everything is movable. Travellers would feel the First Café very interesting.

1

2

第一咖啡位于葡萄牙的老里斯本机场,总面积170平方米。它位于行李大厅的中央,为机场饥饿的旅客提供咖啡和吐司。

先前,这一空间什么都没有。人们有时可能会在这里等上一个小时来取行李,无事可做。整个大厅里只有金属柱子,仿佛地下室一样。这里是一个典型的过渡空间,与其他类似的机场没什么不同。

项目面临的最大挑战是打造一个能够利用柱子的设计。与此同时,这必须是一个开放式店铺,不能封闭现有空间。佩雷拉·米格尔建筑事务所的设计师的设计利用柱子来储藏材料、作为货架并提供了装饰表面。这些装饰以"空间感"为基础,与葡萄牙文化象征相联系,但是经过了现代创新。最大的矛盾在于:过渡空间内的空间感。

为了突出柱子的设计,设计师在机场内设计了小型销售手推车,用来贩售商品和咖啡。它们是缩小版的柱子,看似所有的柱子都能移动一样。游客将认为第一咖啡十分有趣。

1-2. Food and cashier counter
3. Overall view of the shop from outside
4. Interior view of the shop
1、2. 食品和收银柜台
3. 从外部看店铺全景
4. 店铺室内

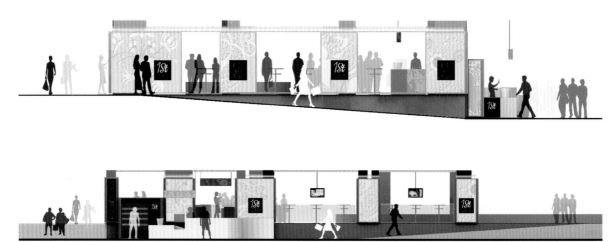

Elevations
立面图

5. Detail of shop display
6. Interior view of the shop
7. Kid area
5. 店铺展示细节
6. 店铺室内
7. 儿童区

Areas Diagram	Circulation Diagram	Product Exhibition, Island, Views Diagram	Ambiences Diagram
区域分析图	路线分析图	商品展示、视角分析图	环境分析图

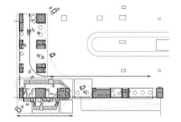 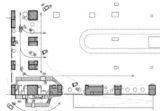 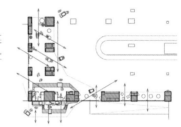 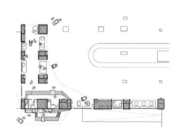

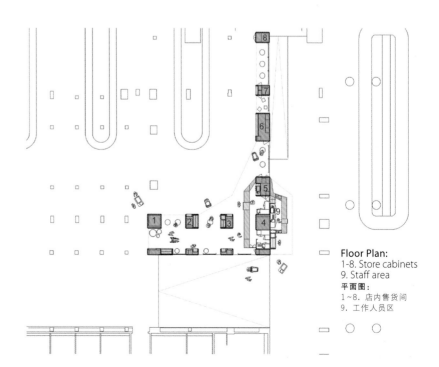

Floor Plan:
1-8. Store cabinets
9. Staff area

平面图：
1~8. 店内售货间
9. 工作人员区

La Maison Des Maitres Chocolatiers
大师巧克力之家

Designer: Minale Design Strategy
Location: Brussels, Belgium
Project area: 55m²
Project year: 2010
Photographs: © Olivier Seignette
设计师：米纳里设计公司
项目地点：比利时，布鲁塞尔
项目面积：55平方米
完成时间：2010年
摄影师：© 奥利维尔·西格尼特

With an ideal location on Brussels' magnificent Grand Place, La Maison des Maitres Chocolatiers Belges offers a unique showcase to ten Master Chocolatiers. Since the opening day the boutique has met with growing success. The board of the Best Belgian Chocolate of the World (BBCW) association is at the origin of this project. They were convinced their ideas deserved an original concept; Minale Design Strategy conducted the design process from start to finish.

The Best Belgian Chocolate of the World Association (BBCW) wanted to promote fine Belgian chocolate by offering a platform to the country's Master Chocolatiers, and create a unique point of sale. "A flagship store concept quickly imposed itself as the best solution", explains Minale Design Strategy Managing Director Gwenael Hanquet. Based on the design that appeals to the emotions, the unique store is designed to welcome

tastings and events. It enshrines the brand and confers visibility to the concept. The Master Chocolatiers now have an exceptional showcase and BBCW has gained a new and effective distribution channel.

1. Package design
2. Storefront design
3. Overall view of interior display
4. Work station
5. Display shelves

1. 包装设计
2. 店面设计
3. 室内展示全景
4. 工作台
5. 展示架

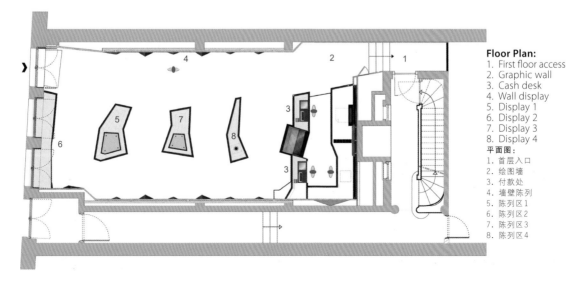

Floor Plan:
1. First floor access
2. Graphic wall
3. Cash desk
4. Wall display
5. Display 1
6. Display 2
7. Display 3
8. Display 4

平面图：
1. 首层入口
2. 绘图墙
3. 付款处
4. 墙壁陈列
5. 陈列区1
6. 陈列区2
7. 陈列区3
8. 陈列区4

6. Chocolate display counter
7. Chocolate display
8-10. Interior design detail
6. 巧克力展示柜台
7. 巧克力展示
8~10. 室内设计细节

大师巧克力之家坐落在布鲁塞尔大广场上,地理位置极佳,以其独特的方式展示着十家大师级巧克力制造商的产品。自开业以来,这家精品店不断获得成功。比利时最佳巧克力协会是这一项目的发起人。他们认为这一概念需要一个原创设计,米纳里设计公司负责项目从始至终的设计。

比利时最佳巧克力协会想要通过提供一个大师级巧克力制造商的展示平台来提升比利时巧克力的形象,同时也打造一个独特的销售地点。米纳里设计公司的总监格温尼尔·汉奇特立刻提出了旗舰店的理念。店铺设计令人愉悦,欢迎人们前来品尝并举办活动。大师巧克力之家崇尚品牌形象,并将此融入设计之中。现在,大师巧克力之家拥有了一个非凡的展示平台,而比利时最佳巧克力协会也得到了一个全新而高效的分销渠道。

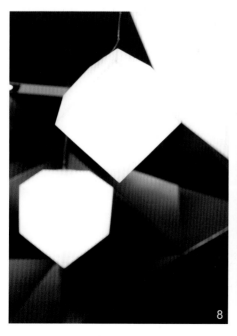
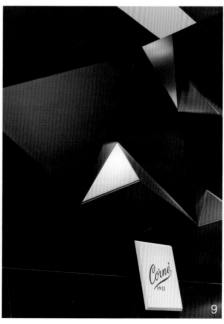

Sprinkles Ice Cream
Sprinkles 冰淇淋

Designer: a l m project INC (Andrea Lenardin Madden)
Location: Beverly Hills, California, USA
Project area: 68m²
Project year: 2012
Photographs: © Trevor Dixon and Andrea Lenardin Madden

设计师： a l m工程设计公司（安德里亚·莱纳尔汀·麦登）
项目地点： 美国，加利福尼亚，比佛利山庄
项目面积： 68平方米
完成时间： 2012年
摄影师： © 特雷弗·狄克逊与安德里亚·莱纳尔汀·麦登

For Sprinkles Ice Cream, the new venture of Sprinkles (known for their cupcakes), it was as important to be identifiable as a new member of the Sprinkles family, as to create a brand that celebrates ice cream, its heritage and iconography.

Exterior like interior derives from a minimalistic design approach and are brought to life by the ample flow of natural light and shadow play. The sparse white façade consists of metal shields with a laser-cut perforation along Sprinkles signature scalloped line applied to the bottom edge. The illuminated cone logo at the top right and a pin-mounted red script "icecream" next to entrance along with the crowd populating the indoor/outdoor bench stretching across almost the entire width of the façade entice the passerby to enter.

Expressing the creamy quality of ice cream, the interior space is defined by the smooth shapes of its main material, white corian, which is employed for both the round counter centre piece and the walls. Accentuated by the pattern of the penny tile base and floor (an analogy to the classic ice cream shop in Europe) and red perforated metal sliders mounted service-side of the display cases, the built-in cabinetry fully integrates both equipment and product.

Highly specific and carefully detailed, the centre piece of the space is further defined by the light rotunda above that features Thomas Jefferson's original ice cream recipe, a metaphor for Sprinkles' approach to make handcrafted American ice cream.

1

3

Sprinkles冰淇淋,Sprinkles公司的新尝试(以他们的纸杯蛋糕而知名),与被识别为Sprinkles家族新的一员同样重要的是,创建一个品牌简单地赞颂冰淇淋,它的传统和它的意象。

与室内相同,室外设计源于一种简约的设计手法,通过充分的光影变幻给设计带来生机。稀疏的白色外观由沿着Sprinkles标识的齿痕线到底部边缘激光穿孔的金属防护罩构成。右上角发光的圆锥体标志和入口旁装饰好的"冰淇淋"徽章与几乎横跨整个外墙宽度的室内外长凳诱惑着路人。

为了表现冰淇淋的奶油特征,内部空间被它的主要材料——白色可丽耐——平滑外表所确定,这种材料被使用在中央的圆形柜台和墙壁上。被底座和地板瓷砖上的便士硬币图案(欧洲的经典冰淇淋店的一个比喻)与安装在展示架服务区一侧的红色金属滑块所凸显,内置的橱柜整合了设备和产品。

高度的独特性和精心细致表现在,空间的中心部分进一步被以汤姆斯·杰斐逊的原始冰淇淋配方为特色的顶部光圆形大厅所定义,这是对Sprinkles的手制美式冰淇淋的一个比喻。

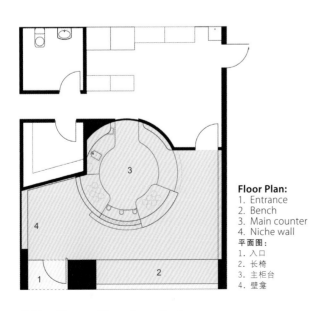

Floor Plan:
1. Entrance
2. Bench
3. Main counter
4. Niche wall

平面图:
1. 入口
2. 长椅
3. 主柜台
4. 壁龛

4   5

1. Package design
2. Entrance design
3. Overall view of the shop
4-5. Service counter details
6. Ceiling design

1. 包装设计
2. 入口处设计
3. 店铺设计全景
4、5. 服务柜台细节
6. 天花板设计

Polka Gelato
波尔卡冰淇淋

Designer: VONSUNG
Location: London, UK
Project area: 32m²
Project year: 2011
Photographs: © VONSUNG
设计师：温桑设计公司
项目地点：英国，伦敦
项目面积：32平方米
完成时间：2011年
摄影师： © 温桑设计公司

VONSUNG recently completed this total identity design for Polka Gelato, from naming, identity, branding, signage, website to spatial design. Based in a conservation area, Fitzroy Square, Polka Gelato opens its doors to showcase their artisanal way of creating ice cream. Despite all the talk of a double-dip recession in the UK, the client's wish was to offer something enlightening, from old to young, a sense of affordable luxury amid these difficult times.

Joseph Sung, of VONSUNG, has strived in his precedent projects to experiment variant ways to explore materials. Among the natural, old, and time-proven material, Sung has derived at lime concrete for this project. Being situated in a historical setting, Sung felt that juxtaposing old and new material would give expected meaning for both, as exemplified using external architectural material within the interior space of the gelato store. Stemming from the client brief, Sung identified with the key word, "artisan", and made every effort not to allow the solid masses of concrete material to feel uncomfortable for the visitors, but feel a sense of skill, the artistry of the space.

The boundaries of the interior wall and ceiling were made to be permeable as possible by way of shadow gaps and openings. Also, to reduce the monolithic manner of concrete, Sung mixed limestones into the batch and applied a smooth finish to the raw concrete. The result was an interior space, which kindles the feeling of being an insider in an environment; simply put, it recognises what may feel like being within a creamy gelato batch. By adopting this method of design, Sung drew the attention to the timeliness of the space and architecture. All faculties of perception and senses, particularly tactility, facilitate the customer experience.

温桑设计公司新近为波尔卡冰淇淋进行了完整的品牌形象设计，从命名、定位、品牌化、引导标示、网站到空间设计，一应俱全。店铺以菲茨罗伊广场为支点，向人们展示着冰淇淋的手工制作工艺。尽管英国正遭受着双重衰退，委托人希望能够提供一些富有灵感的东西，为各个年龄层的人们提供经济实惠的奢华品牌。

温桑设计公司的约瑟夫·沈普在之前设计的项目中试验了各种各样的材料，他在这一项目中采用了自然、古朴而且经得起时间考验的石灰混凝土。店铺坐落在一座历史建筑中，约瑟夫认为将新旧材料并置能够实现两者最佳的预期效果，于是他在室内设计中采用了室外材料。结果便是凸显了项目的关键词——"手工"，并且尽量让顾客在混凝土材料中间感到舒适，体会到空间的技巧和工艺。室内墙壁和天花板的界限通过阴影缺口和开窗变得模糊。为了削减混凝土的块状感，约瑟夫将石灰石分批处理，并且在混凝土的表面进行了光洁装饰。简单来说，置身室内就宛如置身于冰淇淋的世界。通过这一设计方式，设计师强调了空间与建筑的时间线。顾客的各种感官，特别是触感，将增添他们的体验。

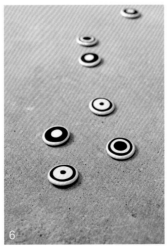

1-2. Storefront details
3. Signage design
4. Interior view
5-7. Logo design
8-9. Dining area

1、2. 店面细节
3. 标牌设计
4. 室内设计
5~7. 标志设计
8、9. 用餐区

Floor Plan:
1. Entrance
2. Shop area
3. Food display
4. Shelf
5. Cash
6. Yard
7. Toilet

平面图：
1. 入口
2. 店铺区
3. 食品陈列区
4. 货架
5. 付款处
6. 院子
7. 卫生间

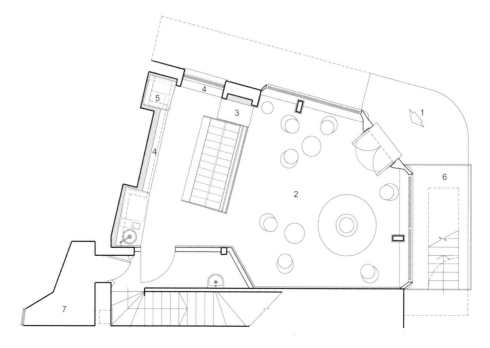

Sweet Chill
甜冰蛋糕店

Designer: Karim Rashid
Location: Las Vegas, USA
Project area: 375m² (Dining: 250m², Kitchen: 125m²)
Project year: 2009.11 (Design period: 2007.4-2008.12)
Capacity: 64 seats for dining & for bar / lounge

设计师：卡里姆・拉希德
项目地点：美国，拉斯维加斯
项目面积：375平方米
完成时间：2009年11月（设计时间：2007年4月~2008年12月）
容量：64人

Like a swirl of ice cream, the MGM Sweet Chill is a continuous gesture throughout the space, with flowing waves from floor to wall to ceiling. Colourful chairs and tables are perfectly aligned, complementing these ribbons and creating a more distinguished circulation. The curvy and sensuous space becomes a natural-looking yet distinctive extension of its surroundings. A continuous clear glass case displaying gelato and paninis grounds the entrance, and menu board spans the back wall creating a focal point. The large columns are covered in chrome, reflecting the surroundings and adding to the overall experience of a completely soft and human environment. The butterfly chair seating, designed by Karim is in his typical bright colours of pink, orange, cyan, and bright lime to maintain the color palate and give variety, and the curvaceous forms fit into the surrounding wave of the wall and ceiling

Floor Plan:
1. Counter
2. Gelato display
3. Point of sale
4. Food preparation
5. Dining space

平面图：
1. 柜台
2. 冰淇淋展示
3. 售货处
4. 食品准备区
5. 用餐区

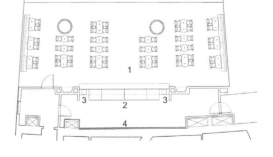

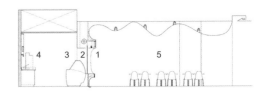

perfectly.

Furniture & Materials:

Flooring: Tile floor by Gastone, partially made from recycled materials, the vinyl which has a low VOC adhesive

Walls & Ceiling: Custom pink curved walls

Dining Chairs: Magis Butterfly Chair

Tables: Custom table by Gist

甜冰蛋糕店在造型上犹如冰淇凌一般，地面、墙壁与天花形成一个连续的整体。彩色的桌椅整齐地摆放着，与带状造型相互呼应，同时使得空间流线格外清晰。蜿蜒的空间造型犹如周围环境的延展，看起来格外自然。玻璃柜台内摆放着冰淇凌和三明治，菜单板横跨正面后墙成为空间的焦点。大柱子采用铬材质覆面，将周围的景象映射出来，同时使得空间看起来更加柔和。设计师亲自打造的蝴蝶座椅色彩多样，粉色、橙色、青色、亮灰色，一应俱全，营造多样性。同时，弯曲的造型则与墙面及天花形成完美的呼应。

家具及材质：

地面：瓷砖、回收材质、低VOC黏胶剂乙烯

墙面与天花：定制粉色弯曲结构

座椅：Magis蝴蝶椅

桌子：Gist定制

1-2. Dining area design
3. Overall view of the shop design
1、2. 用餐区设计
3. 店铺设计全景

Café Bourgeois
资产阶级咖啡店

Designer: Minifie van Schaik Architects
Location: Victoria, Australia
Project year: 2010
Photographs: © Albert Comper
Article: Justin Clemens
设计师：米尼菲·范·斯海克建筑事务所
项目地点：澳大利亚，维多利亚
完成时间：2010年
摄影师：© 阿尔伯特·康珀
文章作者：澳贾斯 · 丁克莱门斯

An idealised three dimensional space is flattened to the wall of the cafe. Café Bourgeois is located in the suburban docklands area, adjacent to the front door of the new Myers building. The café identity and architecture are entwined, the spatial gymnastics implied by the wallpaper mirroring the ironic ambitions of the franchise name.

This idealised three dimensional space imprints on the café walls another worldly quality - a backdrop for glowing displays, the relentless warmth of the polished joinery end-grain and the resilient reception of the rich Marmoleum colours underfoot. It's a highly economical solution, too, a simple sticky-back vinyl print. Yet the striking photographs don't do the subtlety of the interior justice. The juxtapositions effected by the ripple unifying the diverse zones of the café, from the dangling hula-hoop spangled with tiny LED string lights to the display cabinets without overwhelming the space.

One inherent difficulty of designing such spaces requires enhancing function without obtrusiveness: nobody wants to feel too-obviously directed this way or that.

一个理想化的三维空间平铺在了咖啡店的墙壁上。资产阶级咖啡店位于郊外的码头区，毗邻新迈尔斯大楼的正门。咖啡店的特性与建筑交织在一起，墙纸所隐含的空间反映了品牌名称背后具有讽刺意味的野心。

这个理想化的三维空间对咖啡店的墙壁产生了重大的影响，达到了超凡脱俗的品质——一个能够照亮展示柜的背景，抛光的端面纹理的持久温暖和可恢复的接待区脚下的斑斓色彩。一个简单的背黏性乙烯版画，也是一种非常经济的方案。然而，引人注目的图片并不是这一设计的精妙之处。并列的波纹统一了咖啡店的不同区域，从晃来晃去、闪闪发光的LED灯到展示柜，而不会产生铺天盖地的感觉。

设计这种空间的一个固有难题是增强其功能性而不会过于显眼：没有人会愿意感觉他们是被太过明显地引导这么做。

1. Entrance area design
2. Wall detail
3. Food service counter
4. Lighting design of the dining area
5-6. Dining area

1. 入口区设计
2. 墙面细节
3. 食品柜台设计
4. 用餐区灯光设计
5、6. 用餐区

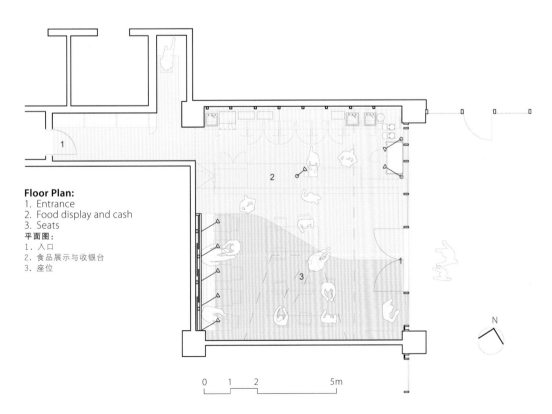

Floor Plan:
1. Entrance
2. Food display and cash
3. Seats

平面图：
1. 入口
2. 食品展示与收银台
3. 座位

Olo Yogurt Studio
Olo酸奶工作室

Designer: Baker Architecture + Design
Location: Albuquerque, New Mexico, USA
Project area: 118.45m²
Project year: 2010
Photographs: © Richard Nunez
设计师：贝克建筑设计事务所
项目地点：美国，新墨西哥州，阿尔奎基
项目面积：118.45平方米
完成时间：2010年
摄影师：© 理查德·努涅斯

The designers were tasked with developing the brand identity, including logo design and demographic targeting, in addition to full architectural services for a new startup in Albuquerque called Olo Yogurt Studio.

Despite the small scale and limited budget, the designers recognised that many people will interact with this space and opportunities for architectural experimentation were as available here as anywhere.

The self-serve yogurt concept calls for a particular circulation pattern in which customers cycle through the space from the dispensers, to the toppings, to the cashier. To draw customers in from the street, the design team developed the "yo-bow" idea which is a series of coloured stripes that undulate overhead, turn downward, and terminate at the yogurt selection at the back of the narrow space, leading customers to the starting point of the assembly line. The primary materials are paint and drywall; the budget was met.

Site Features

The site of this studio is a narrow storefront on the ground floor of a new building in one of Albuquerque's most pedestrian neighbourhoods. Most customers are casual passerbys and residents of the nearby neighbourhoods.

1. Night view of the storefront
2. Food service counter design
3. Overall view of the shop design
4. Cash desk and dining area

1. 店面夜景
2. 食品服务柜台设计
3. 店铺设计全景
4. 收银台和用餐区

设计任务是除了为在阿尔奎基新启动的Olo酸奶工作室提供完整的建筑服务外，还需要打造品牌的身份，包括标志的设计和目标人群。

尽管规模不大预算有限，但设计师认识到许多人将与这个空间互动，和其他任何地方一样，在这里同样有建筑设计的机遇。

自助式的酸奶概念倡导着一种特殊的循环模式，在这里，顾客循环地通过自动售卖机，从配料处，再到收银台，如此往返于其中。为了招揽顾客，设计团队发明了"yo-bow"的点子，那就是在狭小空间的后面一系列的彩条在头顶飘动，向下转，到酸奶选购区结束，引导顾客回到生产线的源头。主要材料是油漆和石膏板，预算得到了满足。

地点特征

这个设计是位于一座新建大厦一层的一个狭小的店面，地处阿尔奎基一个最正宗的步行区内。大部分的顾客是偶然路过的路人和附近的住户。

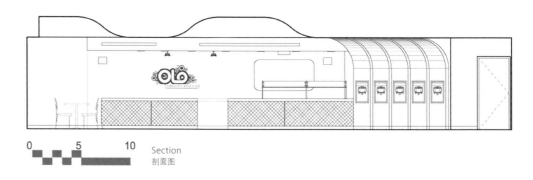

Section
剖面图

5-6. Self-serve yogurt area
7. Dining area
5、6. 自助式酸奶选购区
7. 用餐区

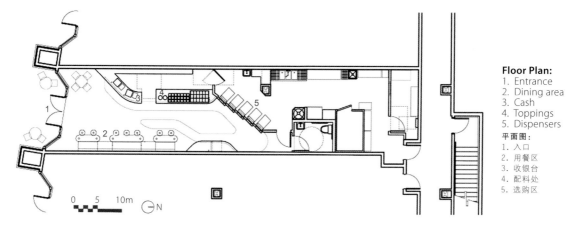

Floor Plan:
1. Entrance
2. Dining area
3. Cash
4. Toppings
5. Dispensers

平面图：
1. 入口
2. 用餐区
3. 收银台
4. 配料处
5. 选购区

Melt Me
融化我巧克力和冰淇淋店

Designer: Mooof by Sunthorn Keeratayakom
Location: Bangkok, Thailand
Project area: 240m²
Project year: 2010
Photographs: © Space Shift Studio
设计师：莫夫工作室（桑松・奇拉塔亚科姆）
项目地点：泰国，曼谷
项目面积：240平方米
完成时间：2010年
摄影师：　© 空间转换工作室

Melt Me is Hokkaido Chocolate & Healthy Gelato shop. It is the first brand in Thailand which is handmade fresh chocolate with Hokkaido style. There are 5 branches in Bangkok and a new store in Pattaya.

The concept design started with chocolate feeling, happiness, cheer, and festival. The colour of Melt Me corporate identity is brown and orange. The orange ribbon wrapping from space to space, created the dynamic and exciting. For the first branch at Thonglor 10, prime area in Bangkok, Mooof designs both architecture and interior space, therefore the main idea is clearly dominated by 2 simple glass boxes, seen through inside and melted with 2 big existing trees.

3

4

融化我是一家巧克力和健康冰淇淋店。它是泰国首家北海道风格的手工鲜巧克力店。融化我在曼谷有五间分店,在芭提雅也有一间分店。

设计理念源自巧克力带给人们的感觉——幸福、喜悦和庆祝。融化我的企业形象以棕色和橙色为主色调。橙色的缎带包裹着各个空间,营造了充满活力和兴奋感的空间。在曼谷的第一家分店设计中,莫夫工作室负责建筑和室内空间设计,采用两个简洁的玻璃盒结构,人们可以一眼看穿整个店铺。店铺还与场地上原有的两棵大树融合在一起。

1. Façade
2. Night view
3. Exterior view from street side
4. Outdoor dining area
5. Food service counter
6-7. Dining area

1. 外立面
2. 夜景
3. 从街道看外观设计
4. 室外用餐区
5. 食品服务柜台
6、7. 用餐区

Floor Plan:
1. Entrance
2. Worktop choc
3. Ice cream
4. Cashier

平面图:
1. 入口
2. 巧克力操作面
3. 冰淇淋
4. 付款区

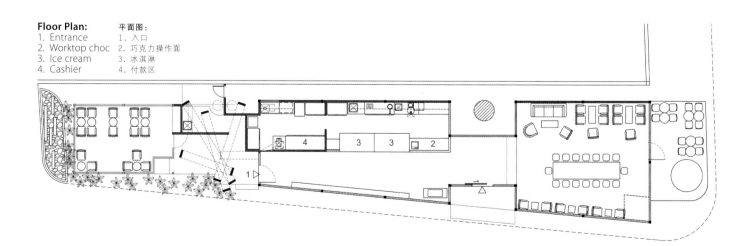

第三章 空间设计

Pusateri's
普萨特里食品店

Designer: GH+A, Lily Yuan
Location: Toronto, Canada
Project area: 915m²
Project year: 2010
Photographs: © Philip Castleton,
Philip Castleton Photography Inc., Toronto, ON
设计师: GH+A，袁荔
项目地点: 加拿大，多伦多
项目面积: 915平方米
完成时间: 2010年
摄影师: © 菲利普·卡斯尔顿；菲利普·卡斯尔顿摄影公司

The concept of "Food is Fashion" was a guiding principle in the design philosophy where food is presented within a visually uncluttered and refined setting. The textures, colours and beauty of the food are the centre of attraction. The fashion look is expressed through the highly reflective surfaces, an aesthetic often found in high-end department stores. Pusateri's feels thoroughly modern and reflects a contemporary lifestyle experience.

Merchandise is integrated into the storefront design in the form of a fully operational cheese fridge that displays large wheels of Parmagiano-Reggiano cheese. Despite the plethora of product offerings, a pared down merchandising programme features product presentations that are structured and contained. Throughout the white space, the different merchandise zones are distinguished according to their materials. The elegance of the design is enhanced with the monolithic use of luxurious materials that form the backdrops for the various departments. All walls are either treated with glass, deeply veined natural stone or darkly stained wood. The flooring throughout the sales area is a large format mini-crystal white tile which is quite daring for a grocery store. White Calacatta marble slabs in a book-matched pattern with concealed overhead lighting are set behind the meat counter. Green glass panels with a customized circular pattern applied on an adhesive film on the back of the glass form the fish department background.

Product displays, though dense, are immaculately organised and the elegant materials are given the prime exposure with their large surfaces. The check-out area is meant to emulate a hotel concierge desk with marble slab desks and none of the traditional check-out cash numbers one would expect to find in a grocery store. In keeping with the store's clean aesthetic philosophy, signage and graphics are minimal.

"食品即时尚"的概念是食品店设计的指导原则。食品的质地、色彩和美感具有核心吸引力。时尚的外观通过高光材质（这种美学设计在高端百货店中十分常见）得到体现。普萨特里食品店反映了现代感十足的时尚体验。

巨型奶酪冷柜嵌入店面设计，展示著名的意大利车轮形帕尔马干酪。尽管店内拥有琳琅满目的产品，精简与归类是设计主旨。在白色为主调的空间内，不同的产品区域由不同装修材质加以区分。整块奢华材料构成的背景墙，体现了优雅的设计。它们或为玻璃幕墙，或为天然纹理大理石，或为优质深色木料。地面采用了大面积的白水晶地砖，突破食品零售店的常规做法。肉类柜台后方，华贵的意大利白色卡拉卡塔大理石板与隐藏式顶灯相互搭配。绿色玻璃板配以圆形图案贴膜，形成了鱼类销售区的背景。

产品展示虽然密集，但是布局精美；同时大面积使用的华贵材料亦成为视觉中心。收银区模仿了酒店礼宾桌，采用大理石台面，丝毫没有传统零售店柜台的气息。为了与店内简洁的美学一致，招牌标识和图案设计也体现了极简主义特征。

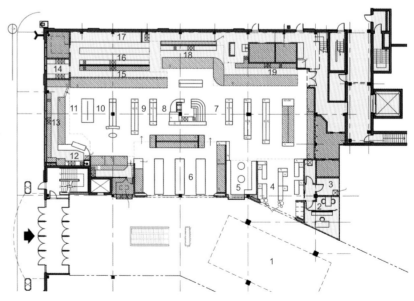

Floor Plan:
1. Seating area
2. Office
3. Cash office
4. Cash area
5. Bread
6. Produce
7. Sushi
8. Demo
9. Coffee
10. Olive oil
11. Chocolate
12. Café
13. Pastry
14. Sink
15. Deli/sandwich
16. Kitchen
17. Cooking with hood
18. Fish
19. Meat

平面图：
1. 座位区
2. 办公室
3. 收银办公室
4. 收银区
5. 面包
6. 制作区
7. 寿司
8. 演示
9. 咖啡
10. 橄榄油
11. 巧克力
12. 咖啡馆
13. 糕点
14. 水槽
15. 熟食店/三明治
16. 厨房
17. 烹饪防护罩
18. 鱼类
19. 肉类

1-2. Storefront design
3-5: Food display design
6. Cash area

1、2. 店面设计
3~5. 食品展示设计
6. 收银区

Raoul's Hammersmith Grove
铁匠林拉乌尔熟食店

Designer: Project Orange
Location: London, UK
Project area: 210m²
Project year: 2011
Photographs: © Gareth Gardner
设计师：橙色项目公司
项目地点：英国，伦敦
项目面积：210平方米
完成时间：2011年
摄影师：© 加雷斯·加德纳

Project Orange has collaborated for the fifth time with Raoul's on their new restaurant and delicatessen in Hammersmith Grove, London. The menu as ever is fantastic home cooked food overseen by proprietor, Geraldine Leventis, presented and served in an elegant environment.

The Deli is a glamorous room with striking Herringbone marble floors, black-mirrored walls juxtaposed with natural oak and marble surfaces. The counters are clad in bespoke woven Lloyd Loom panels, and the room is lit by quirky gold pendants.

Next door the restaurant comprises booths, banquette seating, freestanding tables and a bar. The walls feature a hand-painted signature Harlequin graphic together with black mirror and white buttoned tiles.

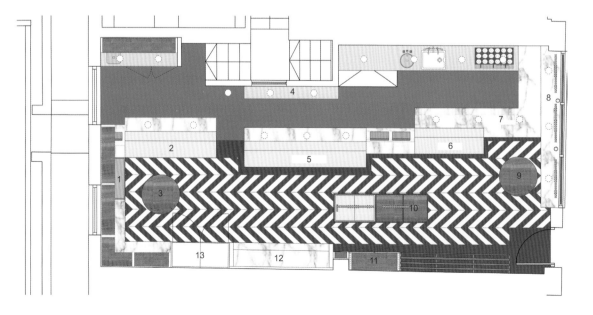

Floor Plan:
1. Cheese display unit
2. Meat/Cheese fridge
3. Display table
4. Worktop
5. Ready meal fridge
6. Cake fridge
7. Pastry display
8. Window display
9. Biscuits display table
10. Bread display
11. Wine display
12. Chilled display unit
13. Entrance

平面图:
1. 奶酪陈列区
2. 肉类/干酪冷藏柜
3. 商品展示桌
4. 厨房操作面
5. 备餐冷藏柜
6. 糕点冷藏柜
7. 面食陈列区
8. 橱窗陈列
9. 糖果饼干陈列桌
10. 面包陈列区
11. 酒品陈列
12. 冷冻品陈列
13. 入口

078 _Chapter Three: Space Design

这是橙色项目公司与拉乌尔的第五次合作，为它在伦敦的铁匠林打造新餐厅和熟食店。店主杰拉尔丁·里文蒂斯在高雅的环境中为顾客提供一如既往的美味家居食品。

熟食店极富魅力，采用非凡的人字斜纹大理石地面、黑色镜面墙壁和天然枫木橡木进行装饰。柜台外包裹着特别定制的织纹面板，整个空间采用奇特的金色吊灯进行照明。

隔壁的餐厅由包间、卡座、独立餐桌和吧台组成。墙壁以手绘滑稽角色图案为特色，采用黑色镜面和白色纽扣瓷砖。

1-3. Food display design
4-5. Dining area design
6. Dining area detail
1~3. 食品展示设计
4、5. 用餐区设计
6. 用餐区细节

Pastry Store "Martesana"
糕点店 "Martesana"

Designer: STUDIO RE: Alberto Re, in collaboration with Silvia Cirabolini
Location: Milan, Italy
Project area: 300m²
Project year: 2010
Photographs: © Diego Decio, Studio Re, Marco Vertua, Silvia Cirabolini
设计师：STUDIO RE 事务所阿尔伯托·雷与西尔维娅·希拉波利尼合作
项目地点：意大利，米兰
项目面积：300平方米
完成时间：2010年
摄影师：© 迭戈·德西奥，Studio Re事务所，马克·威尔多瓦，西尔维娅·希拉波利尼

The intervention was requested by one of the oldest pastries in Milan, which had the need to double the size of the shop. The property has met with courage and enthusiasm the proposal to renew the old image of traditional confectionery, through a project that propose a new meaning and a new way of displaying products: only one level to avoid different plans and then hierarchies of importance.

The materials used (tops in black Corian and not steel as usual) allow a clean image to strongly increase the visibility of goods through the cancellation of an exaggerated expressiveness of materials: cakes are now displayed in glass cases as jewels in a jewellery.

Clean lines and hidden technology: the cooling components disappear in favour of the exhibition of the product, the air vents are reduced to cracks in Corian, the food protecting glass seems suspended being without visible attachments.

Lighting comes from the ceiling (not from the inside of refrigerating aisles) with luminaires never used before in the context of food; the scales are hidden under mount in black glass.

The rest of the store follows the same design principles of showcases: the floor is made of reinforced resin, one-piece without joints visible, some back painted glass plates, vertically installed and three meters high, separate the sales area from the consummation area, and are transformed into exposure hosting products that do not require refrigeration. The rear bench in black lacquered wood stands as the separation of the working area and the sales area, also becoming a full height exhibition area. At the sides there are stairs and the lift to go downstairs, which houses the toilets, stores and workshop of 100m².

All the furnishings are on design: from the signs up to the trays designed in black plexiglass.

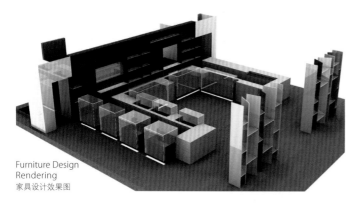

Furniture Design Rendering
家具设计效果图

1. Display shelves   1. 展示架
2. Dining area   2. 用餐区
3-5. Interior details   3~5. 室内细节
6. Dining area   6. 用餐区

该项目所设计的客户是米兰最古老的糕点店之一,他们要求将店铺的面积扩大一倍。这个店铺面临着一个充满勇气与热情的设计方案,用以改变传统糕点店的旧形象,通过这个项目来表现一个新的意义和新的陈列方式:只用一层,而不是多层和强调层次感。

选用的材料(黑色可丽耐而不是通常的钢材)大大提升了食品可见度,而并没有用夸张的、富有表现力的材料:蛋糕就像珠宝一样陈列在玻璃柜子中。

简洁的线条与隐藏技术:为了有力于展示,冷藏设备被隐藏了起来,通风口被设计在可丽耐的裂缝处,保护食品的玻璃罩看起来是悬浮的,并且没有可见的部件。

灯光来自天花板(而不是来自直冷过道的内部),这种灯具之前从未在食品店中使用;秤被隐藏在黑色的玻璃台面下。

店铺的其他部分也遵循着陈列柜的设计原则:地板由无接缝可见的增强树脂组成,后部是3米高、垂直安装的彩绘玻璃板,把制作区和销售区隔开,并把不需要冷藏的货物展示出来。黑色漆木的后排桌椅分隔了工作区和售货区,成为了一个全高的展示区域。旁边是楼梯和电梯,里面安放着洗手间、商店和100平方米的作坊。

所有的家具都经过设计:从标识到托盘都加入了有机玻璃的元素。

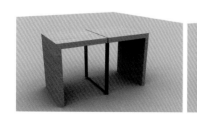 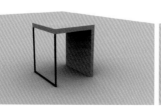  

Table Design 桌子设计

7-9. Food display designs
7~9. 食品展示设计

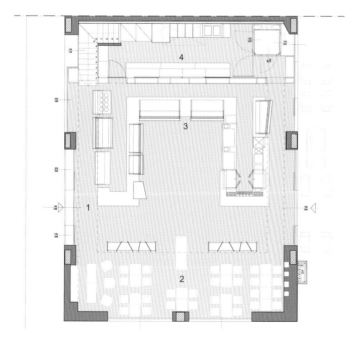
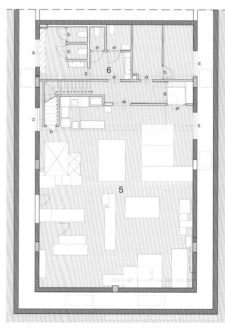

Ground Floor Plan (Left) and Basement Plan (Right):
1. Entrance
2. Dining area
3. Display counter
4. Kitchen
5. Work area
6. Bathroom

一层平面图（左图）和地下室平面图（右图）：
1. 入口
2. 用餐区
3. 展示柜台
4. 厨房
5. 工作区
6. 卫生间

TSUKIAGE-AN
月扬庵

Designer: Aiji Inoue/ DOYLE COLLECTION CO.,LTD.
Location: Kagoshima, Japan
Project area: 81.54m²
Project year: 2012
Photographs: © Satoru Umetsu/ Nacasa&Partners
设计师：井上 爱之/Doyle Collection事务所
项目地点：日本，鹿儿岛
项目面积：81.54平方米
完成时间：2012年
摄影师： © 梅津 聪/纳卡萨摄影公司

TSUKIAGE-AN is Japan's traditional famous shop that processes seafood called "Satsuma age" (a deep fried fishcake). This shop locates in the center of Kagoshima-ken, Tenmonkan, which is a sightseeing spot where tourists from home and abroad gather.

The design of this shop is Japanese modern style. As for the façade, gate shaped frame surround the outer periphery. By adding some lights, it can make differentiation from the other neighboring shops and make it easier to generate original perspective of the world. The designer laid out big display space to express the four seasons, for Satsuma age's flavor and type vary according to each season.

One of the characteristics of the shop's interior design is the louver in the ceiling. Up from the middle top, slanted ceiling adds up the impression of "Wa" (Japanese style). The continuity of the louver has the helpful effect for leading the customers farther into the shop. What's more, there is another characteristic in this shop. You can see how "Satsuma age" are processed in the kitchen. This gives a sense of excitement. You will visually notice that very fresh seafoods are used here. All of the tables and furniture are made full sized mock-up. The designers carefully examined the size, height, visibility of the products in detail. All are custom-made and are designed for easy use.

The shop's logo was used as a hint to make original patterned cloth to garnish the wall surface of the displays. The materials are woven by the same technique as the kimono's "Obi". This will give both sophisticated and splendid impressions.

Floor Plan:
1. Entrance
2. Season's display space
3. Kitchen
4. Showcase
5. Refrigerated display case
6. Waiting
7. Storehouse
8. Toilet

平面图：
1. 入口
2. 本季展示空间
3. 厨房
4. 展示柜
5. 冷藏展示柜
6. 等候区
7. 存储间
8. 卫生间

1. Night view of the storefront
2-3. Overall view of the interior
4. Food display detail
5. Logo design

1. 店面夜景
2、3. 室内设计全景
4. 食品展示细节
5. 标识设计

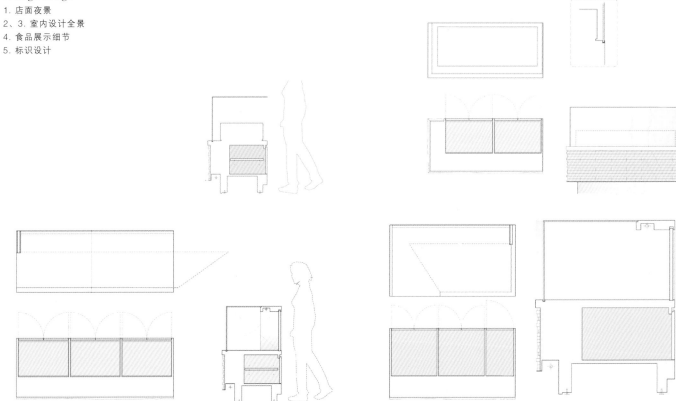

Food Counter Details 食品柜台细节

088 _Chapter Three: Space Design

月扬庵是日本传统名店，生产叫做"萨摩时代"的海鲜（一种油炸鱼饼）。该店位于鹿儿岛县的中心，天文馆，这里是一个吸引国内外游客的观光地。

这家店的设计采用了日本的现代风格。外立面由门形框架包围四周组成。通过增加一些灯光，与其他周边店铺明显区分并增加了原始风貌。设计师使用大的展示空间表达四季，根据萨摩时代的味道和类型，每个季节有所不同。

店铺的室内设计特点之一是天花板上的百叶窗。从顶部的中间开始，倾斜的天花板增添了日式风格。连续的百叶窗设计有助于引领顾客进入店铺深处。

更重要的是，这家店铺有另一个设计特点：你可以看到"萨摩时代"在厨房里如何制作。这是一个很令人兴奋的体验。你会亲眼看到这里应用的是非常新鲜的海产品。所有的桌子和家具都制作了全尺寸的模型。设计师仔细地测量了尺寸、高度、商品的可视度等细节。所有的家具都采用定制并设计的易于使用。

店铺的标识是由传统图案的布料编制而成，它被用来作为墙面的主要装饰元素。这种材料的编制方法与和服的腰带相同——这将会带来一种既复杂又漂亮的感觉。

Blè Food Hall
布莱食品店

Designer: 1+1=1 Claudio Silvestrin, Giuliana Salmaso
Location: Thessaloniki, Greece
Project area: 250m²
Project year: 2011
Photographs: © Yiorgis Yerolymbos
设计师：1+1=1克劳迪奥·希尔维斯特林；朱莉安娜·萨尔玛索
项目地点：希腊，塞萨洛尼基
项目面积：250平方米
完成时间：2011年
摄影师：© 约尔吉斯·耶鲁里默波斯

The interior design for Blè food hall is centred on the idea of the uniqueness of the majestic fire oven.

This extraordinary silent feature, visible from the street, is meant to impress and to make the visitor to ponder about one of the key topics of our time: the source of quality and healthy food.

As per 1+1=1 style, this relevant concept is expressed in a rather poetic way: the fire oven monolithic form, built of natural clay and whitish earth pigment, is expressed as a symbol of fire, earth and sky.

Fire in the earth as that of a volcano, of an organic form that leads our head and eye to move upwards the sky thus showing the impressive 12-metre-high space of Ble' 3 storey volume. Reddish stone on floor and walls adds warmth to the entire space.

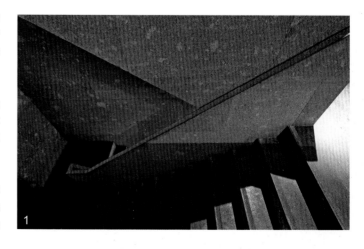
1

1. Staircase details
2. Display and cashier counter
3. Seats detail
4-5. Food display design
1. 楼梯细节
2. 展示与收银柜台
3. 用餐区细节
4、5. 食品展示设计

布莱食品店设计以独特而宏大的烤炉为中心。

从街面上就能感受得到的异常安静的氛围是为了让顾客沉思我们时代的主题：品质的来源和健康食品。

在1+1=1的设计中，这个相关主题以诗意的方式表达出来：巨大的烤炉由天然黏土和发白的土质材料建成，呈现出火、土和天空的主题。

正如火山一样，土中之火的有机形状吸引我们向上仰望天空，展示出布莱食品店12米高（三层）的空间。地面和墙壁上的红石为整个空间增添了一丝暖意。

6. Cakes display
7. Interior lighting design
6. 糕点展示
7. 室内灯光设计

Floor Plan:
1. Oven
2. Bread display
3. Parma ham display
4. Wine display
5. Counter
6. Bread production counter
7. Water feature
8. Window display counter
9. Display feature at column
10. Back area

平面图：
1. 炉灶
2. 面包陈列
3. 火腿陈列
4. 酒品陈列
5. 商品陈列柜台
6. 面包产品柜台
7. 水景装饰
8. 橱窗展示柜台
9. 圆柱陈列
10. 商店后侧

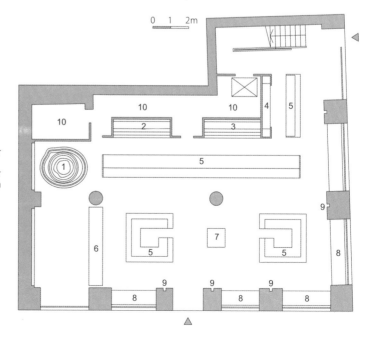

UHA Mikakuto
悠哈糖果店

Designer: GLAMOROUS Co., Ltd.
Location: Shanghai, China
Project year: 2008
Photographs: © Nacasa & Partners, Inc.

设计师： 魅力设计公司
项目地点： 中国，上海
完成时间： 2008年
摄影师： © 纳卡萨摄影公司

The client, UHA Mikakuto Co., Ltd., a famous long-established sweets manufacture in Japan, established its first flag shop in Shanghai, located at a popular streets' intersection.

The shop doubling as showroom has a dominating presence with the continuous huge shades above the showcases and the custom-made lighting stands are covered with two-colour crystals which have been designed and are based on candies and chocolates. While these striking lighting stands catch passengers' eyes, the classical wooden moldings dyed in chocolate colour successfully soften the whole space, being well-balanced. The custom-made lighting fixtures above the long marble table by the window, generating different rhythm to the space, enhance their presence at night looking like floating in the air.

Simple colouration of the interior flatters colourful sweets packages. Within the combined workshop and showroom, "the existence of" concept demonstrated the Major League colour. Custom lighting position is characterised by the size of the gap between deviation paste full colour crystal.

Crystal is also the image of the candy and chocolate. Although the strong impression that the molding made of chocolate stained wood, the general sense of the traditional, classic heavy, balance and overall impression of the software space, maintain a heavy feeling.

Floor Plan:
1. Entrance
2. Cagidereves area
3. R-counter
4. Cosmetic area
5. Stock room
6. Pantry
7. Male's toilet
8. Female's toilet
9. Meeting room
10. Smoking room
11. Seating area
12. Void

平面图：
1. 入口
2. 陈列区
3. 陈列柜
4. 化妆品区
5. 商品展览室
6. 食品室
7. 男士洗手间
8. 女士洗手间
9. 会议室
10. 吸烟室
11. 就餐区
12. 上空

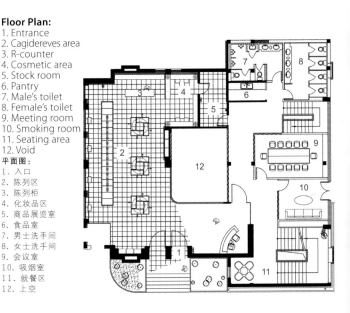

项目的委托人悠哈糖果公司是日本一家驰名的糖果制造商。这是悠哈公司在上海的第一家旗舰店，位于一处繁忙的十字路口边。

糖果店兼具陈列室功能，在陈列柜的上方设有巨大的灯罩，特别定制的灯具采用双色水晶，以糖果和巧克力的色彩为基础。这些非凡的灯具吸引了过往行人的眼球，而经典的木制缘线被漆成了巧克力色，成功地柔化了整个空间，形成了良好的平衡。定制灯具安装在靠窗的大理石长桌的上方，为空间打造了不同的韵律，在夜晚提升了空间的表现力，看起来像浮在空中一样。

室内空间简单的色调凸显了五颜六色的糖果包装。空间结合了工坊和陈列室，"存在感"理念体现了空间的主要色彩。定制灯具以全色水晶为特色。

水晶也体现了糖果和巧克力的形象。巧克力色木制缘线形成了强烈的视觉效果，整体空间显得传统、经典，营造出厚重感。

1. Interior view of the shop
2. Product display detail
3. Display and special lighting design
1. 店铺室内
2. 商品展示细节
3. 展示设计与特殊照明设计

100 _Chapter Three: Space Design

4-6. Product display design
7. Display and special lighting design
4~6. 商品展示设计
7. 展示设计与特殊照明设计

Snog Chelsea
斯诺格冰品店切尔西分店

Designer: Cinimod Studio
Location: London, UK
Project year: 2011
Photographs: © Cinimod Studio
设计师：希尼莫德工作室
项目地点：英国，伦敦
完成时间：2011年
摄影师：© 希尼莫德工作室

The design brief was to create a new high-end interior environment that was a progressive step forward upon the previous award-winnings designs of the previous Snog stores also designed by Cinimod Studio. The Chelsea location, at 155 King's Road, presented unique design challenges that included a deep shop unit with a relatively narrow shop front. This led to the Cinimod design team seeking new and innovative ways to create a highly visible store that would deliver an eye-catching high street presence using a fusion between sculptured architectural and lighting elements.

The original architectural design concept the designers created for Snog evolved around the idea of a "perpetual British summer" – hence the grass floors, floral graphics and digital sky ceilings. In the design studio they have recently been using generative and parametric design tools to assist them in the pursuit of fluid and sensual forms, and at the Chelsea location the designers used this arsenal of tools to full effect to create a seemingly infinitely curving and undulating canopy of "sky ribbons".

Complementing the architectural and lighting design, the designers also developed the main seating element within the store which, unsurprisingly, received the same design treatment, becoming a super-organic form that from some angles looks like the fallen branch of a tree, while from other angles has design references taken from the world of fluid dynamics found in boat and car design.

1. Interior view from the outside
2. Storefront design
3. Overall view of the shop

1. 从室外看室内设计
2. 店面设计
3. 店铺内部全景

设计要求打造一个全新的高端室内环境，延续之前斯诺格冰品店诸多获奖设计的风格。切尔西分店位于国王路155号，其设计挑战在于纵深过深和相对较窄的店面。这驱使着希尼莫德的设计团队探索创新，运用富有雕塑感的建筑和灯光元素打造一个吸引眼球的店面。

设计师为斯诺格冰品店打造的原始建筑设计概念围绕着"永恒的英伦之夏"这一理念展开，拥有草地地板、花朵图案和数码天空屋顶。希尼莫德工作室新近采用了生成参数设计工具来辅助设计流畅而具有感官效果的造型。在切尔西分店的设计中，设计师利用这一工具打造了无穷无尽的曲线和波浪起伏的"天空缎带"穹顶。

为了完善建筑和灯光设计，设计师还在店内利用同样的设计方式开发了主要座椅元素。这些超有机造型从一些角度看来就像掉落的树枝，而有些角度看来又像从船舶和汽车的流线设计中所获得的灵感。

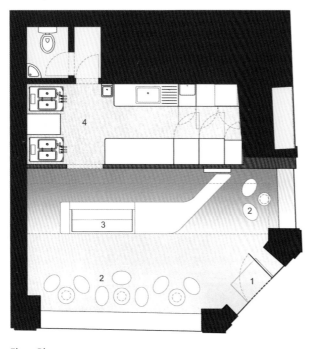

4. Seating area
5. Wall painting
6-7. Interior details
8-9. Details of ceiling
4. 用餐区
5. 墙画
6、7.室内细节
8、9.天花板细节

Floor Plan: 平面图：
1. Entrance   1. 入口
2. Seating area   2. 就餐座椅区
3. Reception   3. 接待区
4. Work space   4. 工作间

第三章 空间设计_ 107

Nana's Green Tea ARIO Kurashiki
七叶和茶仓敷店

Designer: Masahiro Yoshida, Kamitopen Architecture-Design Office Co., Ltd.
Location: Tokyo, Japan
Project year: 2011
Photographs: © Keisuke Miyamoto
设计师：吉田昌弘；吉田昌弘建筑设计公司
项目地点：日本，东京
完成时间：2011年
摄影师：© 宫本橘

Nanaha Corporation is a company that introduces "a new form of Japan" to the world through "matcha" (powdered green tea). They arrange high grade matchas in modern style and provide them as a menu such as matcha latte. And their intention is to make the space into a "contemporary style tea room", not a "Japanese style tea room". This is a reflection of the owner's idea of wanting to make the shops where people can enjoy the traditional Japanese tea culture through a modern concept.

In Okayama Prefecure, there is Korakuen, one of the Japan's three most beautiful gardens. Ikeda Tsunamasa, the second lord of Okayama-han, ordered his men to create Korakuen with the intention of making it a relaxing place. It is a garden that is designed to enjoy the view from the various buildings that are built inside. The scenery has been changed due to the preferences of the lords of each period as well as the situation of the society, but those changes have made the history of the garden.

Normally, a borrowed-scenery would be put in front of the garden. However, the view from Korakuen's tea room includes the Sou Mountain, which is located outside of the garden, making it look like it is part of the garden. As a result of the mountains overlapping with the bright view or the garden, it seems as if garden itself has a depth. Therefore, the designer had suggested to bring in a borrowed-scenery of mountains in the shop. The mountains that come out from the floor function as partitions. And the partitions that hang from the ceiling, which cut the air, are expressing the sky that had been cut by the mountains. By putting both together, the designer has created "a space between the mountains and the sky".

1

奈叶公司向全世界传递"新日本文化",致力于经营高品质的抹茶(绿茶粉)。他们将高品质抹茶以现代方式进行诠释,提供"抹茶拿铁"等新鲜品种。店铺空间并不是典型的日式效果,而是"现代茶室"风格。这反映了店主希望人们能在现代概念下享受日本茶文化的愿望。

冈山市的后乐园是日本三座最美的花园之一。池田光政是冈山的第二任君主,他下令建造了用于放松身心的后乐园。这座花园让人可以享受园内各式建筑的景色。风景随着君主的喜好以及社会的状态而改变,这些改变创造了花园的历史。

通常,借景都设在花园之前。然而,从后乐园的茶室能够看到花园后卖弄苏山的景色。层叠的山峰和花园的景色增添了花园的深度。因此,设计师建议在店铺内采用山峰借景。远处的山峰起到了隔断的作用。而天花板上悬吊的隔断则展现了天空的景色。二者共同营造出"山与天空之间的空间"。

1-2. Interior design details
3. Open storefront design
1、2. 室内设计细节
3. 开放式店面设计

4-5. Dining area designs
4、5. 用餐区设计

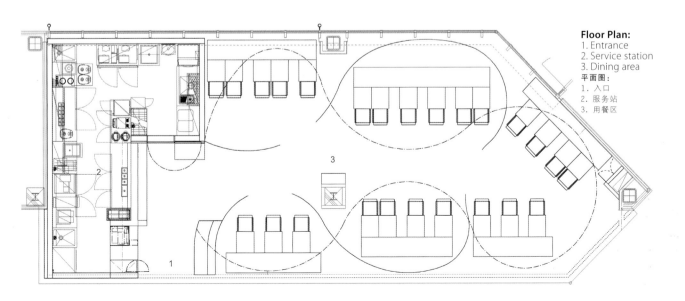

Floor Plan:
1. Entrance
2. Service station
3. Dining area

平面图：
1. 入口
2. 服务站
3. 用餐区

Nana's Green Tea Uehonmachi Yufura
七叶和茶上本町店

Designer: Masahiro Yoshida, Kamitopen Architecture-Design Office Co., Ltd.
Location: Osaka, Japan
Project area: 112m²
Project year: 2010
Photographs: © Keisuke Miyamoto
设计师：吉田昌弘；吉田昌弘建筑设计公司
项目地点：日本，大阪
项目面积：112平方米
完成时间：2010年
摄影师：© 宫本橘

Nanaha Corporation is a company which disseminates a form of "new Japaneseness" to the world, looking at "high quality maccha" (powdered green tea). What is required for the shop is space not in typical Japanese style but a "modern tea room" which is aimed to represent owner's belief that makes customers to enjoy typical Japanese tea ceremony in modern style.

The designers designed "the grating" of the window with a whole new concept. A picture of a tea vine, which is the shop's symbol, comes up and disappears on the grating according to the angle you look at. This means that the designers have added an element of "motion" to a "motionless" interior accessory.

The designers have put tables and chairs that all differ in thickness in order to break down the boundaries of tables and chairs to express a "frank space" where everyone can sit freely at anywhere they wish to. Moreover, the tea quality provides a modern twist on the menu, such as green tea latte.

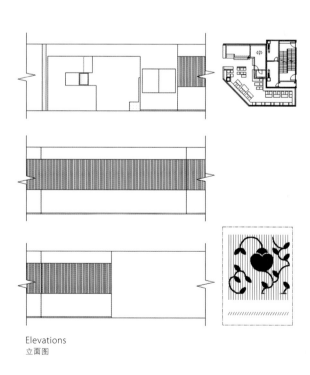

Elevations
立面图

114 _Chapter Three: Space Design

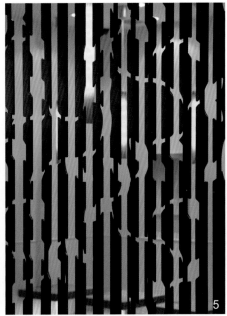
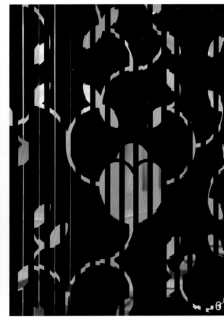

1. Entrance and cashier
2. Exterior view
3. Dining area
4-6. Wall details

1. 入口处与收银台
2. 店铺外观
3. 用餐区
4~6. 墙面细节

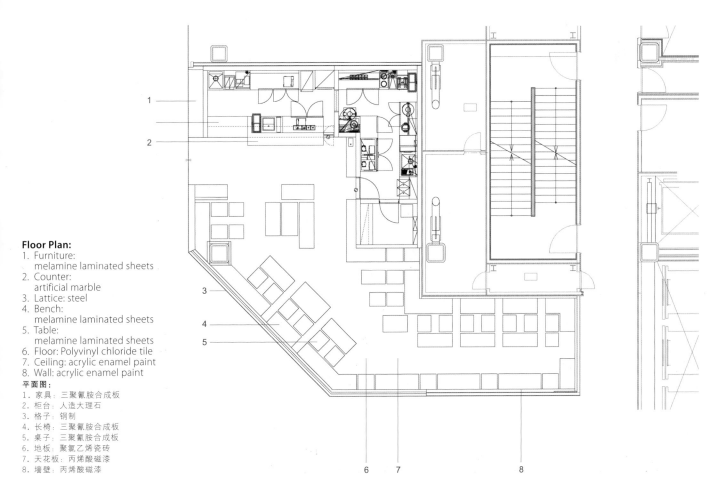

Floor Plan:
1. Furniture: melamine laminated sheets
2. Counter: artificial marble
3. Lattice: steel
4. Bench: melamine laminated sheets
5. Table: melamine laminated sheets
6. Floor: Polyvinyl chloride tile
7. Ceiling: acrylic enamel paint
8. Wall: acrylic enamel paint

平面图：
1. 家具：三聚氰胺合成板
2. 柜台：人造大理石
3. 格子：钢制
4. 长椅：三聚氰胺合成板
5. 桌子：三聚氰胺合成板
6. 地板：聚氯乙烯瓷砖
7. 天花板：丙烯酸磁漆
8. 墙壁：丙烯酸磁漆

118 _Chapter Three: Space Design

7-9. Lighting design of the dining area
7~9. 用餐区灯光设计

奈叶公司向全世界传递"新日本文化",致力于经营高品质抹茶(绿茶粉)。店铺空间并不是典型的日式效果,而是"现代茶室"风格,体现了店主让顾客在现代风格中享受典型日式茶道。

设计师打造了全新理念的"窗格"。茶树藤图案是店铺的象征,在窗格上随着角度不同而呈现出不同的效果。这在静止的室内装饰中增添了"动态"元素。

设计师在室内添加了厚度不同的桌椅,打破了桌椅的界限,呈现出人人能够随意就坐的直率简单的空间。此外,七叶和茶的菜单也变得更为现代,如绿茶拿铁。

Nana's Green Tea Sendai Parco
七叶和茶仙台店

Designer: Masahiro Yoshida, Kamitopen Architecture-Design Office Co., Ltd.
Location: Sendai City, Japan
Project area: 115m²
Project year: 2011
Photographs: © Keisuke Miyamoto
设计师：吉田昌弘；吉田昌弘建筑设计公司
项目地点：日本，仙台
项目面积：115平方米
完成时间：2011年
摄影师： © 宫本橘

Nanaha Corporation is a company that introduces "a new form of Japan" to the world through "matcha" (powdered green tea). They arrange high-grade matchas in modern style and provide them as a menu such as matcha latte. And their intention is to make the space into a "contemporary style tea room", not a "Japanese style tea room". This time the designers created a contemporary "seascape".

Shakkei means to characterise a garden by using natural scenery, such as mountains, woods and bamboo groves, around it. Normally, a view that can be taken in a tea room is shakkei, but from the porch of Kanran-tei, which was a tea room of Hideyoshi Toyotomi at the Castle of Fushimi Momoyama and later moved to Miyagi Prefecture by Masamune Date. We are able to look at the sea of Matsushima. Therefore, the designer has designed Nana's Green Tea in Sendai into a modern tea room where you can view a "seascape". He set an indirect lighting in the middle of the wall surface to resemble the sea horizon and line up logs which differ in diameters and graining in the upper and lower parts in order to describe the depth of sea and sky by graduations of various colours and sizes. Moreover, in the logs placed lower than the sea horizon, the designer gave them a naguri process (a distinct wave pattern) by an adze to express the ripples.

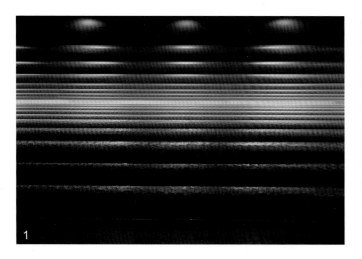

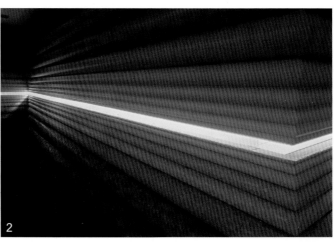

奈叶公司向全世界传递着"新日本文化",它致力于经营高品质的抹茶(绿茶粉)。他们将高端的抹茶产品以现代的方式进行诠释,提供"抹茶拿铁"等新鲜饮料品种。该店铺空间并不是典型的日式室内效果,而是"现代茶室"的风格。设计师将其营造成为一幅现代"海景画"。

"借景"意为利用自然景观来塑造花园,比如周边山、树林和竹林等。通常情况下,茶室的景观都利用借景的手法,但是通过宫城县观澜亭的门廊茶室的观感,我们能够看到松岛的海。因此,设计师将七叶和茶仙台店设计成为一个能够观看海景的现代茶室。

设计师利用墙壁中央的间接照明来模仿海平面的起伏效果,比如用直径、纹理和色彩不同的起伏关系的原木来描绘海洋和天空的高度、深度和远度。此外,"海平面"下的原木被排列成独特的波浪图案,呈现出了"涟漪"的特殊效果。

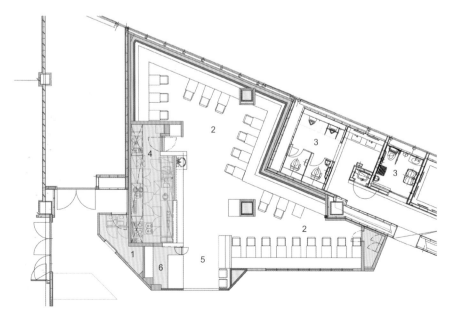

Floor Plan:
1. Stock
2. Seat
3. Restroom
4. Cook
5. Goods counter
6. Reception

平面图:
1. 存储
2. 用餐区
3. 卫生间
4. 食品准备区
5. 商品柜台
6. 接待处

1-2. Wall details
3. Dining area
4. Storefront design
5. Exterior view
6. Dining area
7. Dining area
1、2. 墙面细节
3. 用餐区
4. 店面设计
5. 店铺外观
6. 用餐区
7. 用餐区

Chapter 4: Food Service Counters

第四章：食品服务柜台

Service Stations
服务台

Customer Self-Service Buffets
顾客自助服务餐台

Beverage Service
饮品服务台

130 Espresso Areas
咖啡餐台

131 Alcohol/Bar Facilities
酒品餐台/吧台

131 Vending Machines
自动售卖机

131 Liquid Foods and Ice
液体食品及冰饮

132 Self-Service Beverages
自助服务饮料

132 Beverages in Paper-Based Packaging
纸包装饮料

132 Can Openers/Stirring Mechanisms
开罐器/搅拌设施

Above and Facing above from left: Food service counter design
上图及对页上两图，从左至右：食品柜台设计

FOOD SERVICE COUNTERS

Service Stations
Custom fabricated cabinets used in the wait station, alcohol service area, or customer self-service area must be finished with plastic laminate that meets the local standards. All exposed surfaces of the cabinet(s), including the underside of the cabinet or countertop, must be finished with plastic laminate or equivalent.

Cabinetry may be installed in a food service establishment only in areas approved by the regulatory authority. Cut outs in millwork shall be sealed by the fabricator. All cabinets shall be on six-inch legs or on a solid masonry base with approved base cove installed. Enclosed hollow bases are not permitted. Hand sinks may not be installed in plastic laminated counters except in a limited food service. The bottom shelf under any plumbing or refuse area is recommended to be removed.

食品服务柜台

服务台
等候区、酒品服务区和顾客自助区服务柜台应需采用塑料层板饰面，用以满足相关的食品安全标准等的要求。此外，所有裸露在外的表面结构（橱柜下侧和柜台表面）都必须采用塑料层板或其他相似结构饰面。

在食品店内，只有一些特定的区域可以摆放橱柜。另外，木质橱柜的接缝处需要保持密封状态，所有的橱柜都必须带有离地面高达6英寸的柜脚或者采用固体砖石的底座（不可以采用空心的底座，处于卫生等方面的考虑），除非经过特殊批准，否则洗手盆不可以安装在塑料层板的柜台上，管道或是废物区的底架建议移除。

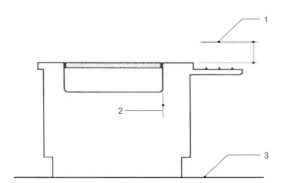

Cooking Food Counter :
1. Bottom leading edge of foodshield
2. Front inside edge of displayed food
3. Finished floor

烹调食品柜台：
1. 食品防护罩的底部前缘
2. 展示食品的前内侧边缘
3. 完成的地面

CHAPTER ONE FOOD SERVICE COUNTERS 第四章 食品服务柜台

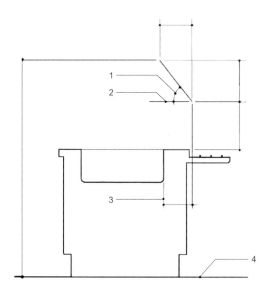

Self Service Food Counter:
1. Angle varies
2. Bottom leading edge of foodshield
3. Front inside edge of displayed food
4. Finished floor

自助食品柜台：
1. 角度可以变化
2. 食品防护罩的底部前缘
3. 展示食品的前内侧边缘
4. 完成的地面

Customer Self-Service Buffets

Customer self-service buffets shall utilize mechanical refrigeration and/or hot-holding units, be indirectly wasted to a floor drain, be located on a smooth, durable, easily cleanable floor which extends three feet beyond the edge of the salad bars and buffets.

Single-service articles shall be dispensed individually wrapped or from an approved dispenser. Cabinetry must be constructed to local standards with plastic laminate on all exposed surfaces. Wood cabinetry is not permitted. Sealed wood cabinet doors on breakfast buffets may be allowed.

Approval prior to installation is required: cabinetry must be installed on with six (6) inch legs or a solid masonry base with a cove base installed at the base/floor juncture; countertops must be either solid surface, granite, stainless steel or other approved material(Laminate countertops are not permitted); approved food shields must be provided.

Beverage Service
Espresso Areas
A separate hand washing sink may be required and the hand sink may not be used as a dump sink. A separate dump sink may be required. A running water dipper well or a means for supplying clean working utensils must be provided. A knock box for emptying coffee must be provided. Trash cans may not be used as a knock box.

顾客自助服务餐台

顾客自助服务区餐台应该配备冷藏柜和保温柜，并且要直接与地面的排污系统连通，更要放置在光滑、耐磨并且易清洁的地面上（确保地面应多出餐台边缘 3 英寸）。

单独供应的食品需要确保包装完好或者放置在特定的容器内加以保护。橱柜的表面应采用塑料层板装饰，并确保不能使用木质橱柜（早餐台柜门可采用木材打造）。

橱柜安装之前需要满足如下条件：所有的橱柜都必须带有高达 6 英寸的柜脚或固体砖石底座；餐台表面须采用固体材质、大理石、不锈钢或其他材质（不可采用层板结构）装饰；食品必须进行遮盖，确保安全卫生。

饮品服务台
咖啡餐台
咖啡餐台需要配备独立的供员工或顾客的洗手盆以及独立的垃圾清理池（两者不可混淆）。确保使用流动的水流用于清洗设备或其洗涤用途，台面上应该配备咖啡敲渣盒（不可以用垃圾桶代替），用于清理咖啡渣。

Above: Espresso area design
上图：咖啡区设计

Alcohol/Bar Facilities

At least one hand washing sink is required depending on size (more hand sinks may be required). A separate dump sink shall be provided. If a four-compartment sink is used, the first compartment may be used as the dump sink. All equipment, including refrigeration must meet local standards.

Wood used as the bar top should be a hard-wood, such as maple or oak, and be finished with a minimum of three coats of polyurethane or equivalent. Stone or tile finished bar tops shall have an approved sealer applied so that the surface is impervious to liquids and grease. Other materials must be submitted to the Regulatory Authority for approved prior to installation.

Vending Machines

Vending machines, although technically regarded as "food shops", often do not require the same level of construction and equipment as full-fledged food shops. They do have some specific requirements to ensure the safe storage and dispensing of food and the prevention of health hazards.

Liquid Foods and Ice

For vending machines that dispense liquid food or ice, it is important to prevent the entry of condensate or splash, which may be contaminated, into the food product. Food contact surfaces which divert liquid food into the receiving container need to be protected from contact by customers/people to prevent contamination of the food product. A self-closing door

酒品餐台 / 吧台

酒品餐台应至少配备一个洗手盆和一个独立的污水排水池。如果安装的是四格水池，那么第一格应用做污水排放。所有的设备包括冷藏柜等都必须符合相关标准。

吧台面如使用木材装饰，则必须采用硬木材质，如枫木或者橡木等，并采用聚氨酯等材质饰面。吧台面如果采用石材或瓷砖装饰，则应确保接缝处的黏合，以免液体或油渍等的侵蚀。吧台面如果采用其他材质装饰，则必须符合相关要求。

自动售卖机

自动售卖机，虽然被视作食品店柜台的一种，但其设计通常不要求完全遵守食品店的相关规定。其本身需要特殊的标准以确保食品在储藏和配发过程中的卫生安全。

液体食品及冰饮

对于配发各类液体食品以及冰饮的自动售卖机来说，防止机器的冷凝和喷洒至关重要，以免污染到里面的食品。盛放食品的容器表面应该有所保护以避免与顾客或其他人直接接触，防止食品污染或导致浪费。同时，对于放置在室外的或无人监管的自

Above and Facing above from Top: Food service counter designs

on outdoor machines or unsupervised machines further protects against accidental or malicious contamination.

Self-Service Beverages

Self-service beverage dispensing equipment should be designed to prevent contact between the lip-contact surface of glasses or cups that are refilled and the dispensing equipment actuating lever or mechanism or the filling device. Beverage equipment that utilizes carbonation equipment (CO_2) shall incorporate a back-flow, back-siphonage prevention device (i.e., check valves) to prevent the migration of the carbonated beverage into copper water supply lines.

Beverages in Paper-Based Packaging

Vending machines designed to store beverages that are packaged in containers made from paper products should be equipped with diversion devices and retention pans or drains for container leakage.

Can Openers/ Stirring Mechanisms

Cutting and piercing parts of can openers on vending machines must be protected from manual contact, dust, pests, and other contamination. Both openers and stirring mechanisms must be cleaned on a regular schedule. Cutting and piercing parts of can openers on vending machines come into direct contact with the canned food product and, if not protected, may contaminate the vended food product.

动售卖机应该配备自动关闭门，确保意外不发生和不受到污染。

自助服务饮料

自助服务饮料配发装置应该被设计成防止杯体与嘴唇接触和由于续杯带来的污染以避免盛放容器与饮料杯口与出水结构的直接接触。配备碳化设备的自助服务饮料装置应该采用防倒流和"反虹吸"设计（即单向阀），防止碳酸饮料进入到饮用水供应管道中去。

纸包装饮料

用于存放纸包装饮料的自动售卖机内，应该配备有分水装置和排水设施两种，用来解决容器的泄漏等问题。

开罐器／搅拌设施

由于自动售卖机上的开罐器的切割和穿孔设备要与灌装食品直接接触，因此，防止灰尘、昆虫等的侵蚀尤为重要，确保卫生安全。开罐器和搅拌设施必须定期清理，如果不加以保护，可能会污染售卖机内的食品。

第四章 食品服务柜台

Oliver Brown
奥利弗·布朗食品店

Designer: Morris Selvatico
Location: Sydney, Australia
Project year: 2010
Photographs: © Ben Cole
设计师：莫里斯·塞尔瓦迪克
项目地点：澳大利亚，悉尼
完成时间：2010年
摄影师： © 莫里斯·塞尔瓦迪克

Oliver Brown located in Chatswood Westfield's recently refurbished food precinct provided an ideal surrounding for the brand's second store. Amongst the hustle and bustle of grocery shopping Oliver Brown provides a haven for chocolate and coffee lovers alike to take time out and just relax.

In order to capture the essence of the authentic and timeless qualities inherent to the art of chocolate making, the designers opted to reflect a vintage feel in the design. This is not only reflected in the interior but in the branding and packaging as well. A balance of bold patterns, aged timbers, handcrafted chocolates and an inviting ambiance sets this café concept apart from its neighbouring competition and gives Oliver Brown its design point of difference.

The rustic terracotta flooring and recycled spotted gum timber resides alongside the bold wallpaper graphic, whose geometric forms relate directly to the very roots of the word "chocolate" and its Aztec heritage. It is this pattern that goes on to inform the packaging and branding of the Oliver Brown chocolate products.

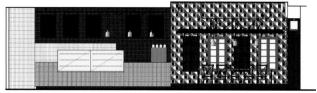 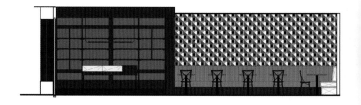

Elevations 立面图

1

2

1. Cashier counter
2. Storefront
3. Overall view of interior
4. Dining area detail

1. 付款台和陈列柜
2. 店面
3. 店内全景
4. 用餐区细节

奥利弗·布朗食品店位于查兹伍德·西野商场新近翻修的食品区域,是该品牌的第二家店铺。在熙熙攘攘的食品店中间,奥利弗·布朗食品店为巧克力和咖啡爱好者提供了休闲的天堂。

为了捕捉巧克力制作艺术真实而经典的品质,设计师选择在设计中反映复古感。这不仅体现在室内设计,而且还体现在品牌形象和包装设计上。大胆的图案、陈旧的木料、手工巧克力和诱人的氛围让这家咖啡馆远离周边店铺的竞争,为奥利弗·布朗提供了与众不同的设计出发点。

纯朴的陶土色地面和回收的斑点桉木与大胆的壁纸图案相配合。壁纸图案与"巧克力"这个词的来源和巧克力的阿芝特克(墨西哥的印第安原住民)历史紧密相连。这种图案同时还出现在奥利弗·布朗巧克力产品的包装和品牌设计上。

Elevation 立面图

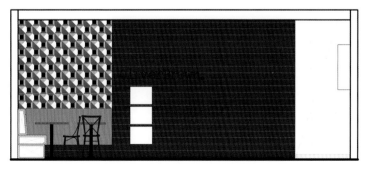

Rear Wall Elevation 后墙立面图

5-6. Display counter
5、6. 展示柜台

Floor Plan:
1. Custom cold display unit with custom stainless steel fridge under
2. Salamander
3. Contact grill (if required)
4. Pos/cash drawer under bench
5. Microwave
6. Chocolate tempering machine
7. Coffee machine
8. Grinder
9. Ice machine (under bench)
10. Single door under bench fridge
11. Hand wash basin
12. 2-door-under-bench freezer
13. Dish washer
14. Single bowl wash up sink with 200mmh splash guard for sink separation
15. Single bowl sink for food prep
16. Spray arm tap
17. Stainless steel bench prep bench with s/steel shelving over allow for s/steel drying rack over wash up sink
18. 4-burner gas cooktop
19. Stainless steel commercial exhaust hood over cooktop
20. Cool room

平面图：
1. 客户定制冷藏展示配客户定制不锈钢电冰箱
2. 面火炉
3. 接触烤架（如需要）
4. 收款机
5. 微波炉
6. 巧克力融化炉
7. 咖啡机
8. 研磨机
9. 制冰机
10. 单门电冰箱
11. 洗手盆
12. 双门冷柜
13. 洗碗机
14. 单边洗碗槽配200毫米高挡板
15. 单边食品准备水池
16. 喷雾式水龙头
17. 不锈钢长椅配不锈钢置物架
18. 4燃气炉灶面
19. 不锈钢商用排气罩
20. 冷室

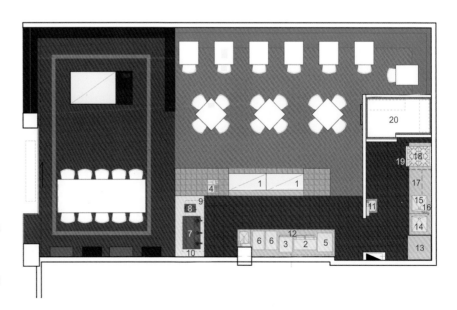

Jewels Artisan Chocolate
珠宝手工巧克力店

Designer: Nota Design International Pte Ltd.
Location: Singapore
Project area: 96m²
Project year: 2009
Photographs: © Nota Design International Pte Ltd.
设计师：诺塔设计国际公司
项目地点：新加坡
项目面积：96平方米
完成时间：2009年
摄影师： © 诺塔设计国际公司

Presented with a design brief to create an up-market shop at Orchard Central, Nota Design International Pte Ltd. has morphed Jewels artisan chocolate's flagship shop into one that stimulates visitors' senses of taste and eyesight as much as their chocolates presented in delicious colours and decorations.

To design an up-market chocolate boutique café in a shopping mall located at Orchard Road, the brief derived from the name "Jewels Artisan Chocolate". The design offered the experience of viewing and selecting a crafted jewellery (in this case the chocolate) in this new flagship shop.

The requirement of an enclosed kitchen together with a chocolate making viewing area set the challenge for spatial planning as the mall management insisted that being an island shop space. The new shop implementation must not block the surrounding shops.

The design of the display counters have to be different from the normal counters and have to become a draw in their own rights. The notion of the "uncut gem stones" was adopted in the design of the display counters.

The design of the shop has been well received by the clients and the shoppers. The careful choosen of material used for the display counters, the walls and the ceiling have created an interesting ensemble that not only embodies the concept of crafted jewelleries but also echoes the status (high end image) of the brand.

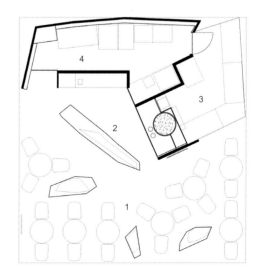

Floor Plan:
1. Café
2. Counter
3. Open concept kitchen
4. Preparation area

平面图：
1. 咖啡区
2. 商品陈列柜
3. 开放式厨房
4. 准备区

1. Display counter
2. Overall view of interior from the outside
3. Detail of storefront
4-5. Seating view
1. 展示柜台
2. 从外部看店内全景
3. 店面细节
4、5. 用餐区

诺塔设计国际公司受委托在乌节中央城打造一个高档店面——珠宝手工巧克力店,利用店面设计与美味可口的巧克力和装饰一起刺激顾客的味觉和视觉。

为了在乌节路一家购物中心打造一家精品巧克力店,设计围绕着店名"珠宝手工巧克力"展开。设计为顾客提升了在旗舰店内体验手工巧克力的感觉。

封闭式厨房和开放式巧克力制作区为空间规划提供了挑战,因为购物中心要求店铺呈岛型设计,并且新店铺不能遮挡周边的店铺。

展柜设计必须与普通柜台相区别,具有独特的吸引力。展柜设计以"未切割的宝石"为基本理念。

店铺设计深受委托人和顾客的喜爱。展柜、墙面和天花板的材料经过精挑细选,营造了有趣的整体效果,体现了手工珠宝的理念,还与品牌的高端形象呼应。

第四章 食品服务柜台_ 143

Little Bean Blue
Little Bean Blue 咖啡店

Designer: Zwei Interiors Architecture
Location: Melbourne, Australia
Project area: 40m²
Project year: 2011
Photographs: © Michael Kai
设计师：Zwei 室内建筑事务所
项目地点：澳大利亚，墨尔本
项目面积：40平方米
完成时间：2011年
摄影师：© 迈克尔·凯

Little Bean Blue is a new concept for an established Melbourne retail coffee trader. The approach is streamlined with the enterprise purely focused on single origin coffee, save for a small selection of sweet treats.

The desire to communicate and educate inspired the project with the design referencing educational spaces such as science laboratory, classroom and lecture theatre. Walls have been transformed into interchangeable layers of blackboards displaying descriptive diagrams that explain various coffee preparation processes, communicating with customers in a unique graphic manner.

Vertically stacked tiles stretching into the space suggest scientific diagrams, as if the production and preparation of the coffee beans is conducted via carefully tested and proven methodology delivering the perfect coffee for the customer.

Simple rendered walls and industrial style feature lighting give materiality to the space, with featured cobalt blue highlights such as the Tolix stools and the bench-top drip tray creating definition and interest. A subtle light-green bulkhead creates contrast to the blackboards and materials below, and recycled timber boards to the rear create a central focal point and warmth to the space.

146 _Chapter Four: Food Service Counters

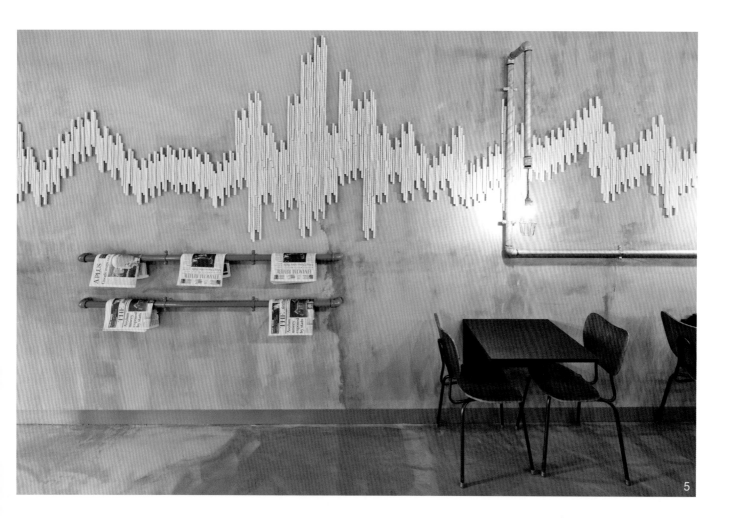

Little Bean Blue咖啡店是墨尔本一个著名的咖啡零售商的一个新概念。此种方法是只为顾客提供单一产地的咖啡和提供一小部分的甜点。

这个项目的设计灵感来源于对沟通和教育的欲望，设计中也参考了科学实验室、教室与演讲厅等的设计。墙面已经被改造成可换层的黑板，上面有描述性的图表来解释不同的咖啡的制备工艺，咖啡馆使用这种独特的图形方式来与顾客进行着沟通。

堆叠并垂直延伸的瓷砖代表着科学性图表，就像是为了给顾客提供完美的咖啡，咖啡豆的生产和准备都经过了仔细的测试和验证。

简单的渲染墙与工业风格照明给空间以物质感，钴蓝色的设计亮点，例如Tolix凳子和台式滴水盘创建了这个设计的定义和兴趣。一个精妙的浅绿色隔板与黑板及下面的材料形成对比，后部再生木材制成的木板创建了一个中央焦点并为空间带来了温暖感。

1. Dining area
2. Entrance and outdoor dining
3. Overall view of the shop
4. Food service counter
5. Dining area

1. 用餐区
2. 入口处与室外用餐区
3. 店铺全景
4. 食品服务柜台
5. 用餐区

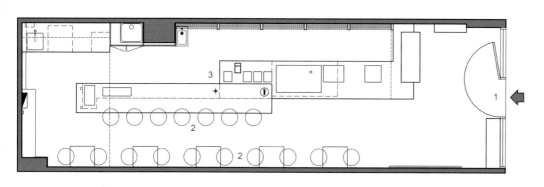

Floor Plan:
1. Entrance
2. Seats
3. Counter

平面图：
1. 入口
2. 座位
3. 柜台

Coffee Hit
Hit咖啡

Designer: Zwei Interiors Architecture
Location: Melbourne, Australia
Project area: 160m²
Project year: 2011
Photographs: © Michael Kai

设计师：Zwei室内建筑事务所
项目地点：澳大利亚，墨尔本
项目面积：160平方米
完成时间：2011年
摄影师： © 迈克尔·凯

The design concept for this tenancy references disused spaces that are reinvented and appropriated for other uses. A large black and white graphic wraps the rear wall with a 1:1 scale image of a decrepit; disused warehouse in a state of decay suggests that the site is unfinished whilst hinting at what could have previously occupied the space.

A striking double-line of white cylindrical pendants stretches from the front of the tenancy, over the large communal table to the rear wall, drawing the customer's eye into the space. Upon entering the rear seating area, a corner wall of floor to ceiling pre-loved books reveals itself as an installation of art sustaining the narrative of adaptation and evolution.

The subtle palette comprising timber, charcoal and white intensifies the red elements with the overall ambience being dark and moody. Coffee Hit creates respite from the over-lit retail shopping mall, allowing for restoration before the continuation of a shopping assignment.

这个项目的设计理念在于改造废弃的空间并做其他用途。一个巨大的、1:1比例制成的黑白破旧图像包裹住了后墙，衰败状态中的废弃仓库表明该地点是未完成的，同时暗示着这一空间之前的用途。

一行双排的圆柱形白色吊灯从店铺的前端，从就餐区的上方延伸到后墙，这一设计吸引了顾客的眼球。走进后部就餐区，会发现墙角从地板到天花板堆满了二手书，这一装置宛如一件艺术品，表现着持续性的适应与进化。

由木材、木炭和白色所构成的精妙色调在黑色、忧郁的主题氛围中凸显了红色的设计元素。Hit咖啡在购物中心里为人们提供了休息的机会，在继续购物前恢复体力。

Floor Plan:
1. Entrance
2. Seats
3. Counter
4. Kitchen

平面图：
1. 入口
2. 座位
3. 柜台
4. 厨房

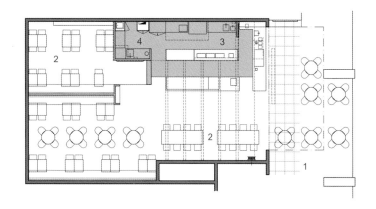

152 _Chapter Four: Food Service Counters

1. Dining area details
2. Overall view
3-4. Dining area
1. 用餐区细节
2. 店铺全景
3、4. 用餐区

Cafenatics
Cafenatics咖啡店

Designer: Zwei Interiors Architecture
Location: Melbourne, Australia
Project area: 100m²
Project year: 2011
Photographs: © Michael Kai
设计师：Zwei室内建筑事务所
项目地点：澳大利亚，墨尔本
项目面积：100平方米
完成时间：2011年
摄影师：© 迈克尔·凯

Located at the rear of a CBD office tower, the design concept was based on the desire to create a retreat from the hustle and bustle of city life. The expresso machine and food servery sit central to the space, acting as a divider to the seating zones and presenting the espresso machine as the heart of the tenancy.

With a palette inspired by a simple country shack, the materials include recycled timber from industrial palettes stacked vertically with openings to create a rustic texture, a stippled black & white horizontal forest graphic stretching behind the servery wall, simple timber chairs and green metal stools. Featured lighting drops from the dark ceiling as exposed florescent tubes, the bright vertical white light aligning to the forested image backdrop.

The mood is both tranquil and intimate, as if retreating to a forested hideaway, to submission to great coffee and food, the antidote to the whirl of contemporary working life.

项目位于中央商务区办公大楼的后面,设计理念是基于远离城市生活的喧嚣这一愿望。咖啡机与上菜区位于空间的中央,分割了就餐区,并使咖啡机处于店铺的中心位置。

店内色调来源于一个乡村的窝棚,工业色调的再生木材从顶端垂直堆叠以表现出质朴的纹理,一个黑白斑点的垂直森林图案包裹住了上菜处后面的墙、简单的木椅子和绿色的金属凳。裸露的荧光灯管构成了这个空间里的特色照明,白色的垂直光束从黑暗的天花板上倾泻下来照射在森林图案的背景上。

店内的氛围宁静而亲切,如同逃离到了一处世外桃源,屈从于美味的咖啡和美妙的食物,是忙碌的现代生活的解毒剂。

1. Food service counter
2. Exterior view of the shop
3. Food service counter
1. 食品服务柜台
2. 店铺外观
3. 食品服务柜台

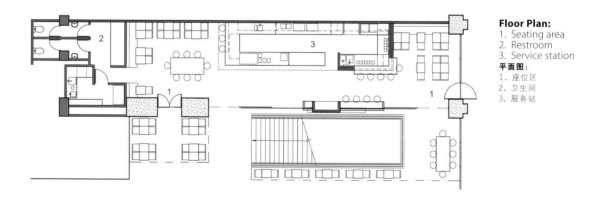

Floor Plan:
1. Seating area
2. Restroom
3. Service station

平面图：
1. 座位区
2. 卫生间
3. 服务站

158 _Chapter Four: Food Service Counters

4. Food service counter
5. Wall painting of the dining area
4. 食品服务柜台
5. 用餐区墙画

Chapter 5: Food Display
第五章：食品展示

Indoor Display
室内展示

162 Display
展示

163 Operation
操作

Outdoor Display
室外展示

Facilities to Maintain Product Temperature
食品温度调控设施

164 Refrigeration Facilities Sizing and Design
冷藏设施规格和设计

164 Hot Holding and Reheating Facilities
保温及加热设备

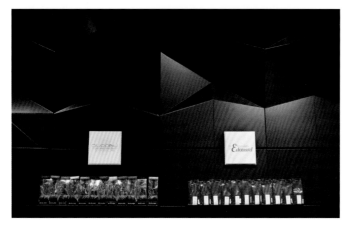 

Above from left: Chocolates display design
上图，从左至右：巧克力展示设计

INDOOR DISPLAY

Display

Except for nuts in the shell and whole raw fruits and vegetables that are intended for hulling, peeling, or washing by the consumer before consumption, food on display shall be protected from contamination by the use of packaging, counter, service line, or sneeze guards that intercept a direct line between the consumer's mouth and the food being displayed, containers with tight-fitting securely attached lids, display cases, mechanical dispensers, or other effective means.

Non-prepackaged food may be displayed and sold in bulk in other than self-service containers if both of the following conditions are satisfied: the food is served by a food employee directly to a consumer; the food is displayed in clean, sanitary, and covered, or otherwise protected, containers.

Raw, non-prepackaged food of animal origin, such as beef, lamb, pork, and eviscerated fish, shall not be offered for consumer self-service. This subdivision does not apply to the following: consumer self-service of ready-to-eat foods at buffets or salad bars that serve foods such as sushi; ready-to-cook individual portions for immediate cooking and consumption on the premises, such as consumer-cooked meats or consumer-selected ingredients for Mongolian barbecue, or raw, frozen shrimp, lobster, finfish, or scallop abductor muscle, or frozen breaded seafood.

室内展示

展示

除了一些带皮坚果、生鲜水果、蔬菜等食物在食用之前需要人们去壳、剥皮或者清洗之外，其他的食品应该采用包装袋包好，用柜台、喷嚏挡等结构在店内展示，主要是用来避免顾客无意触碰到或着随意品尝。盛放食品的容器应该确保为密封状态，也可以采用陈列柜台、机动分配器等其他有效的方式展示。

未经过包装的食品通常采用成块或成批的方式展示或售出，如不采用自助容器，则要求满足以下条件：食品由员工直接递送给顾客；食品盛放在干净卫生的容器中。

未经包装的生鲜肉类比如牛肉、羊肉、猪肉或去内脏鱼类等食品不应该提供顾客自助形式的服务，但以下情况除外：自助餐台或沙拉吧上的即食食品，比如寿司这类食品；店内食用的即时烹调的食品，如顾客自助的肉类、虾类、贝类等食品。

未包装食品采用成块或整批展示和自助售卖，并满足如下条件：深度加工的食品，生鲜肉可放在开放

Above from left: Cakes display design
上图，从左至右：糕点展示设计

Non-prepackaged food may be displayed in bulk for consumer self-service if all of the following conditions are satisfied: produce and food requiring further processing, except raw food of animal origin, may be displayed on open counters or in containers; except for salad bar and buffet-type food service, a label shall be conspicuously displayed in plain view of the consumer and securely attached to each self-service container, or in clear relationship to it, and shall contain the information required in section; non-food items shall be displayed and stored in an area separated from food.

French style, hearth-baked, or hard-crusted loaves and rolls shall be considered properly wrapped if contained in an open-end bag of sufficient size to enclose the loaves or rolls.

Consumer self-service operations for ready-to-eat foods such as buffets and salad bars shall be provided with a suitable food dispensing utensil for each container displayed or effective dispensing methods that protect the food from contamination. Consumer self-service operations such as buffets and salad bars shall be checked periodically on a regular basis by food employees trained in safe operating procedures.

Operation
The dispensing operation is installed contiguous with a permanent food facility and is operated by the food facility. The beverages are dispensed from enclosed equipment that precludes exposure of the

式柜台或容器内；自助餐台和沙拉吧之外的食品应配备标签说明，确保黏合在容器上并向顾客展示信息；非食品类物品应在单独区域存放和展示。

法式烘焙食品（面包或糕饼）如放置在开口的袋子中，应考虑用适当的方式包装，确保其不受损坏。

顾客自助服务的即食食品如自助餐台或沙拉吧上的食品应配备合适的分发容器，便于展示并防止食品受到污染。顾客自助服务设施如自助餐台和沙拉吧需经受过食品安全培训的专业人员定期检查。

操作
店内应配备食品分取设备。饮料要从封闭设备内分取，顾客自助饮料分取设备避免与容器表面接触。

冰块及冷冻食品必须通过专用的分取容器盛放，不可用手直接拿取或放置在室外。

一次性使用应防污染，并放在专用的分取设备中。分取设备上方需一定的遮盖，确保其免受污染。

停止营业的时间内，分区设备应遮盖起来，避免昆虫等其他有害物质的侵蚀。

Above: Candy display design
上图：糖果展示设计

Above: Chocolates display design
上图：巧克力展示设计

beverages until they are dispensed at the nozzles. The dispensing equipment actuating lever or mechanism and filling device of consumer self-service beverage dispensing equipment shall be designed to prevent contact with the lip-contact surface of glasses or cups that are refilled.

Ice and ice product are dispensed only from an ice product dispenser. Ice and ice product are not scooped or manually loaded into a dispenser out-of-doors.

Single-use utensils are protected from contamination and are individually wrapped or dispensed from approved sanitary dispensers. The dispensing operations have overhead protection that fully extends over all equipment associated with the facility.

During non-operating hours the dispensing operations are fully enclosed so as to be protected from contamination by vermin and exposure to the elements.

The permit holder of the permanent food facility demonstrates to the enforcement agency that adequate methods are in place to properly clean and sanitize the beverage dispensing equipment.

Condiments shall be protected from contamination by being kept in dispensers that are designed to provide protection, protected food displays provided with the proper utensils.

OUTDOOR DISPLAY

Only prepackaged non-potentially hazardous food or uncut produce may be displayed or sold outdoors by a food facility if all of the following conditions are satisfied: outdoor displays have overhead protection that

食品店经营者应向相关部门保证具备足够的措施确保饮料分取设备的清洁与卫生安全。

调味品应放置在专用的分取设备内，避免污染，展示中的食品应采用专用的容器盛放。

室外展示

仅有那些不易损坏的包装食品适用于室外展示或出售，但须满足如下条件：室外展示区应具备顶棚遮蔽结构；除营业时间之外，展示食品应在室内存放；室外展示应经许可，并定期检查。

食品温度调控设施

遵循相应标准，要配备足够的冷藏和保温设施。

冷藏设施规格和设计

为满足存贮需求，食品店内应划分足够的空间放置冷藏设施，便于食品存放、运输、展示等。详细的冷藏条件要求需根据食谱、食品移动频率等制定。

食品店内经营潜在危害食品必须配备冷藏和冷冻设施。冷藏制冷组合设施不可放置在烹饪或高温加工设备附近，以免增加制冷负荷。

保温及加热设备

保温设备必须将潜在危害食物在展示和存储过程中的温度保持在140华氏度（60摄氏度）或更高温度。加热设备则必须将潜在危害食物的温度维持在最低165华氏度（约等于73.9摄氏度）。此外，需配备合适的食品温度表以检测温度。

Above: Cupcake display wall
上图：纸杯蛋糕展示墙

Above: Food display detail
上图：食品展示细节

extends over all food items; food items from the outdoor display are stored inside the fully enclosed food facility at all times other than during business hours; outdoor displays are under the control of the permit holder of the fully enclosed food facility and are checked periodically on a regular basis.

FACILITIES TO MAINTAIN PRODUCT TEMPERATURE

Sufficient hot-holding and cold-holding facilities shall comply with the local standards or equivalent, and shall be designed, constructed and installed in conformance with the requirements of these standards.

Refrigeration Facilities Sizing and Design

The plan review for storage needs to provide adequate refrigeration facilities for the proper storage, transportation, display, and service of potentially hazardous foods. Specific refrigeration needs will be based upon the menu, number of meals, frequency of delivery, and preparation in advance of service.

Provide point-of-use refrigerators and freezers at work stations for operations requiring preparation and handling of potentially hazardous foods. Refrigeration units, unless designed for such use, should not be located directly adjacent to cooking equipment or other high heat producing equipment which may tax the cooling system's operation.

Hot Holding and Reheating Facilities

The hot holding facilities must be capable of maintaining potentially hazardous foods at an internal temperature of 140EF or above during display, service and holding periods. Reheating equipment must be capable of raising the internal temperature of potentially hazardous foods rapidly (within a maximum of 2 hours) to at least 165EF. Appropriate product thermometers will be required to monitor the food temperature.

William Curley
威廉姆·科利点心店

Designer: Jonathan Clark Architects
Location: London, UK
Project year: 2009
Photographs: © Richard Dean

设计师：乔纳森·克拉克建筑事务所
项目地点：英国，伦敦
完成时间：2009年
摄影师：© 理查德·迪恩

This is a new shop and dessert bar for renowned chocolatiers William & Suzue Curley – William is the holder of the "Best British Chocolatier" award for 2007, 2008 and 2009 by the Academy of Chocolate. They already have a small shop in Richmond which has received numerous accolades and reviews in the national press.

This project involved the complete strip-out and subsequent fit-out of a 90-square-metre ground floor retail space and 60-square-metre basement. At ground floor level is a 7-metre-long chilled cabinet for displaying fine patisserie, chocolates and ice cream in a temperature controlled environment. Adjacent to this is a dessert bar with seating for 7. There are also three separate areas of banquette seating to give another 16 covers.

One of the main ideas behind the design of the interior was to think about chocolate in terms of its origins and how it is procured – designing a predictable dark brown interior to "represent chocolate" was something the designers actively avoided. This resulted in features such as the gold leaf artwork for the corner banquette depicting cocoa trees and the three-dimensionally arranged hessian clad padded panels behind the counters to reference the hessian sacks that cocoa beans are packed in at various plantations around the world.

The custom-made clear glass lighting pendants to the rear are loosely intended to resemble cocoa pods. The remainder of the palette consists of reclaimed pine flooring, limestone counters and surfaces, burnt orange leather, oak joinery and a long spray-lacquered ceiling raft to unify the various elements, highlight the length of the space and help to draw the eye down to the back of the shop.

At basement level is a demonstration kitchen where chocolate courses and workshops will take place.

这家点心店由知名巧克力制造商威廉姆与苏祖·科利经营——威廉姆在2007年、2008年和2009年连续三年获得巧克力协会所评选的"最佳英国巧克力制造商"。他们在里士满已经拥有一家获得了无数媒体赞美的小店。

项目涉及90平方米的零售空间和60平方米的地下室的全盘设计和后续装饰。一楼有一个7米长的冷柜,用于在温控环境里展示精致的法式糕点、巧克力和冰淇凌。冷柜旁是一个点心吧,可供7人就坐。店内还有三个独立的能够容纳16人的就餐区域。

室内设计的主要理念考虑了巧克力的来源和获得过程,设计师极力避免以深棕色室内来呈现巧克力。因此,在转角的卡座里通过金叶艺术品来描绘出可可树,用立体布局的亚麻布包裹柜台后方的面板来模仿世界各地用来放置可可豆的麻袋。

店铺后部特别定制的玻璃吊灯象征着可可豆的豆荚。其他的设计元素包括回收的松木地板、石灰石柜台和台面、鲜橙色皮革、橡木制品以及长条喷漆天花板。设计师在处理天花板时,将不同的元素统一在一起,突出了空间的长度并将人们的眼球吸引到店铺后部。

地下室主要是用来展示的厨房,里面设置着巧克力制作课程和工坊。

Ground Floor Plan:
1. Entrance
2. Display
3. Corner banquette
4. Front seating
5. Main chocolate counter
6. Till
7. Dessert bar
8. Kitchen
9. Rear seating

一层平面图:
1. 入口
2. 商品陈列
3. 角落软长椅
4. 前端座椅区
5. 主巧克力陈列柜台
6. 铁柜
7. 甜品吧
8. 厨房
9. 后端座椅区

第五章 食品展示_ 169

1. Display counter
2. Drink display and dining area
3-5. Detail of wall painting and ceiling design
6. Interior view of shop
7. Display detail
8. Detail of drink bar

1. 商品陈列柜台
2. 饮品陈列与就餐区
3~5. 墙画细节和天花板设计
6. 店铺内景
7. 商品陈列细节
8. 饮品吧细节

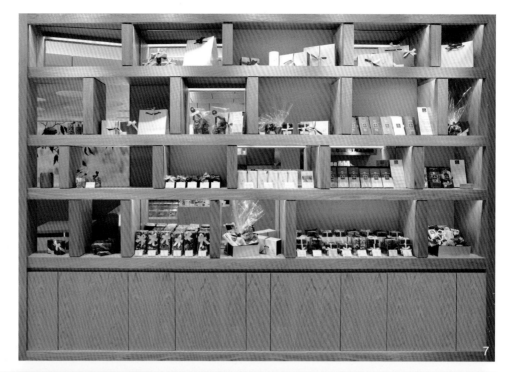

Bea's of Bloomsbury
Bea's of Bloomsbury蛋糕店

Designer: Carbon
Location: London, UK
Project area: 110m²
Project year: 2010
Photographs: © Carbon
设计师：Carbon事务所
项目地点：英国，伦敦
项目面积：110平方米
完成时间：2010年
摄影师：　© Carbon事务所

London based creative + boutique studio, carbon, have recently completed work on the much anticipated 2nd location of Bea's of Bloomsbury; an independent boutique cafe specialising in bespoke cakes and high quality food.

The new location comprises two floors: a kitchen and serving area on the ground floor, and a mezzanine level with bench and booth seating, offering customers a casual relaxed environment in which to look out through the glazed facade at St Paul's. The site is a narrow 70m² corner unit in the newly developed One New Change building by Jean Nouvel adjacent to St. Paul's Cathedral. The design of the restaurant is modern with a twist of the aesthetic mind-set of a Japanese tea house. With much of the food offered at Bea's being cute and colourful, the design approach went for a dark neutral background to celebrate the skilled hands at Bea's. Cupcakes, cakes and quaint objects are displayed in individual glass boxes, showcasing the food as stars of the show.

The upholstery throughout the store compliments the colours of Bea's food with sprinkles of different coloured cushions contrasting with the darker background. The palette was kept tight with dark anthracite on vertical surfaces and soft blue undertones on ceilings and feature areas.

The lighting is kept ambient with a soft comfortable glow from the ceramic teapot lights suspended at various

1. Cupcake display wall
2. Corner perspective
3. Façade elevation
4. Mezzanine and ground spaces
5. Ground level seats
6. Mezzanine level seating and lights
1. 纸杯蛋糕展示墙
2. 街角外观
3. 外观立面
4. 夹层与一层空间
5. 一层用餐区
6. 夹层用餐区与灯光设计

heights, adding a layer of fun and elegance to the scheme. At night, the display boxes and teapot lights output a welcoming glow that puts a smile across the faces of people passing by.

"The focus was in creating a unique and experiential space for Bea that reflected the same care and passion she had for the food she offered. The challenge was in creating an environment that reflected her culture which customers could recognize and engage, despite being an independent retailer." says carbon co-founder/director Go Sugimoto.

Elements of her packaging design, also by carbon, can be seen with large scale graphics swooping through the scheme. The graphic is broken up along the various vertical surfaces of the space, which can be seen in its complete form along the elevation. "In creating a strong and present identity, it is crucial to maintain a design language that is consistent throughout her offer" explains Sugimoto, "the store is her experiential identity."

Front Elevation
前立面图

第五章 食品展示_ 175

坐落在伦敦的创意精品工作室——Carbon，于近日已经完成了期待已久的第二家蛋糕点的设计工作，这是一家专营定制蛋糕和高级美食的独立精品餐厅。这个门店共有两层：厨房和服务区在一楼，楼上是独立的小包间，透过玻璃窗还能看到圣保罗大教堂，在这里顾客可以享受一个休闲轻松的环境。 这是在Jean Nouvel新开发的圣保罗大教堂附近的"One New Change"大厦里一个70平方米的狭窄的角落单元。这个餐厅的设计融合了现代与日本茶室的审美观点。Bea's的设计手法趋于黑暗中立的背景来衬托厨师们制作可爱和丰富多彩的食物的娴熟手艺。茶点、蛋糕和精致的点心都用独立的玻璃盒来展示，就像一场明星盛宴一般。

整个室内装饰和Bea's食物的色彩相得益彰，不同颜色的坐垫与黑暗的背景形成鲜明对比。这种墙壁与地面的黑色背景配以天花板和功能区柔和低调的蓝色所形成的色彩混搭表现很紧凑。

这里的照明与从悬挂在不同高处的陶瓷茶壶上反射的柔和光线融为一体，为方案平添了一份乐趣和优雅。夜晚，展示盒和茶壶的光营造出欢迎光临的氛围，把微笑带给每一个路人。"这个设计理念的重点是营造一个独特的体验空间，正如Bea对待所提供的食物的那份精心和热情一样。而挑战就是要创造一个能反映已被客户认知并参与其中的文化环境，虽然这是一个独立的零售门店。" Carbon创始人及董事Go Sugimoto说。

它的包装设计元素也由Carbon完成，体现在贯穿整个方案的大规模图纸。图纸被分割成了大量的空间垂直平面，从它完整的立视图可以看出来。"为了营造强大和当下身份，保持设计理念与经营产品一致是至关重要的，"Sugimoto解释，"这个门店就是她的体验式身份。"

Ground Floor Plan: 一层平面图：
1. Servery　　1. 上菜处
2. Kitchen　　2. 厨房
3. Toilet　　 3. 卫生间
4. Seating　　4. 座位
5. Display　　5. 展示

Mezzanine Plan: 夹层平面图：
1. Bench seating　1. 长凳区
2. Booth seating　2. 雅座区
3. Servery　　　　3. 上菜处
4. Display　　　　4. 展示
5. Open to below　5. 开向下方

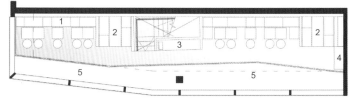

176 _Chapter Five: Food Display

Margaret River Chocolate/ Margaret River Providore
玛格利特河巧克力店/玛格利特河供应店

Designer: Jody D'Arcy
Location: Perth, Australia
Project area: 205m²
Project year: 2012
Photographs: © Jody D'Arcy
设计师：乔迪·达尔西
项目地点：澳大利亚，珀斯
项目面积：205平方米
完成时间：2012年
摄影师： © 乔迪·达尔西

The very popular Margaret River Chocolate Company and Margaret River Providore have opened a new retail store in Perth, Western Australia, within a decorative Edwardian inner city heritage building.

The warehouse type space have been filled with long benches, shelves and wall displays specifically designed to showcase the extensive range of vividly packaged products. The main internal wall has been clad with richly grained timber boards which are then adorned with packaged chocolate bags hanging from oversized timber pegs, all suspended from loosely tied black rope.

Colours, materials and finishes are purposefully limited to ensure the merchandise quality and abundant quantity remain the foremost message. The space is simple, approachable and consistent with the products on display.

180 _Chapter Five: Food Display

1-3. Details of the display design
4. Products display area design
5. Products display area design
1~3. 陈列设计细节
4. 商品陈列区设计
5. 商品陈列区设计

深受欢迎的玛格利特河巧克力公司和玛格利特河供应店在澳大利亚珀斯新开了一家零售店,位于爱德华内城一座精致的历史建筑之内。

仓库式空间内设满了长椅、货架和墙面展示,专门为展示大量包装活泼的商品而设计。主要内墙上覆盖着纹理丰富的木板,上方装饰着挂在木桩上的精美巧克力袋子,所有袋子都以黑色绳子松松地系上。

色彩、材料和装饰的精心选择都保证了商品的质量和数量保持在最佳状态。整个空间简洁、亲切,并且与展示的产品相得益彰。

6-7. Products display area design
6、7. 商品陈列区设计

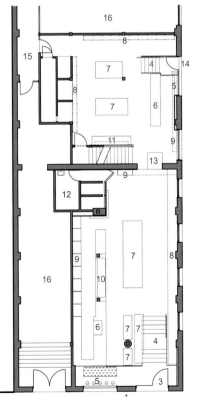

Floor Plan:
1. Murray street footpath
2. Wolf lane
3. Street entry
4. Stairs
5. Window display
6. Service counter
7. Display tables
8. Bench display
9. Display shelving
10. Glass display
11. Cabinet display
12. Store
13. Ramp
14. Lane entry
15. Service corridor
16. Adjoining tenancy

平面图：
1. 莫里街小径
2. 里弄
3. 街边入口
4. 楼梯
5. 橱窗展示
6. 服务台
7. 陈列桌
8. 长椅陈列
9. 陈列架
10. 玻璃陈列架
11. 橱柜陈列
12. 储藏
13. 坡道
14. 里弄侧入口
15. 服务通道
16. 隔壁租赁

Café Chocolat
诱惑巧克力吧

Designer: Studio Linse Amsterdam
Location: Amsterdam, the Netherlands
Project area: 250m²
Project year: 2009
Photographs: © Went & Navarro

设计师：林斯工作室，阿姆斯特丹
项目地点：诱惑巧克力吧
项目面积：250平方米
完成时间：2009年
摄影师：© 温特&纳瓦罗

Seductive chocolate lounge with mouth-watering materials in an organically shaped interior design pleases all senses.

The project aims to translate the seductive chocolate experience into an inspiring and expressive space, challenging all senses and to an attractive retail zone with a comfortable and embracing seating area in an otherwise open and transparent airport location.

The soft touch of chocolate led to a feminine approach, using organic shapes and a strong and decorative pattern echoing the lace-like paper cup supports of the original "patisserie". The purple ribbon overhanging the complete area resembles the chocolate box ribbon, binding the neighbouring Bubbles seafood & wine bar together creating an indulgent zone and giving intimacy tot the seating area.

The project located in the busy commercial area of this airport. Around the corner from the high end shops it is paramount to get attention, be noticed and communicate the purpose of the space without blocking the wonderful view nor the airport logistic flows, providing a welcoming place to relax. The Main display element which accommodates both cash desk and food preparation, is decorated with a lacelike pattern inspired by the paper doily used in the French Patisseries. The large display case holds chocolate treats in well designer gift packaging that travellers can purchase to take home.

To enhance the chocolate experience guests are invited to sniff fragrant glass bottles and to guess the kind of chocolate that produces each aroma.

The seating area on the upper level allows guests to enjoy their induces in a moderate private setting. The tables next to the window provide an interesting view of air traffic. Both tables and condiment holders were custom designed by Studio Linse.

诱人的巧克力吧利用有机造型，室内设计里令人垂涎欲滴的材料愉悦着人们的各个感官。

项目旨在将诱惑的巧克力体验延伸到富有表现力的空间里，挑战人们的感官，在相对开放透明的机场空间里打造一个具有吸引力的零售空间，并配备舒适的休息区。

巧克力的柔和质感吸引着女性，用有机造型和富有装饰感的蕾丝式纸杯来盛放原创的法式糕点。

悬挂在整个区域上方的紫色缎带与巧克力盒上的缎带有异曲同工之妙，将旁边的气泡海鲜酒吧连接在一起，营造了一个宽松的区域，为休息区提供了私密感。

项目位于机场繁忙的商业区。在高端店铺的环绕中，它仍然引人注目，同时又不遮挡视野、妨碍机场人流移动，提供了一个舒适友好的休闲空间。主要展览元素兼具收银台和食品准备功能，采用蕾丝图案进行装饰，其灵感来源于法式糕点的纸巾包装。巨大的展示柜里摆放着精美包装的巧克力，游客可以买一些带回家。

为了增强巧克力趣味体验，店铺邀请顾客嗅一嗅盛在杯中的巧克力来猜测巧克力的种类。

上层休息区让顾客在私密的环境中尽情休息。临窗的桌子提供了机场有趣的交通景象。桌椅和调味瓶都由林斯工作室特别设计。

1. Dining area detail
2. Display counter
3. Overall view of the shop from the outside

1. 餐饮区细节
2. 展示柜台
3. 从室外看店铺内部全景

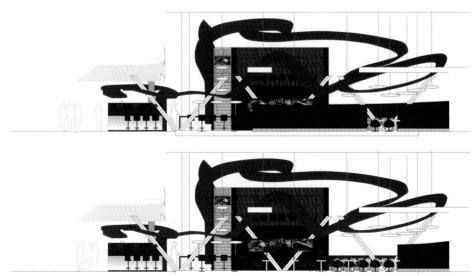

4-5. Dining area
4、5. 餐饮区

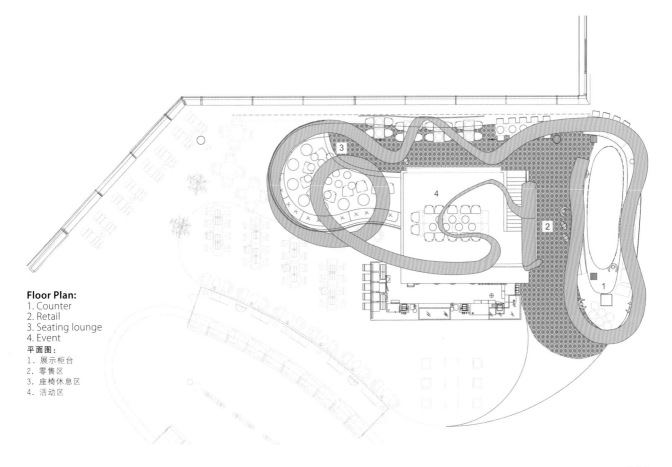

Floor Plan:
1. Counter
2. Retail
3. Seating lounge
4. Event

平面图：
1. 展示柜台
2. 零售区
3. 座椅休息区
4. 活动区

Nascha's
娜莎食品店

Designer: Denis Kosutic
Location: Vienna, Austria
Project area: 100m²
Project year: 2011
Photographs: © Lea Titz

设计师：丹尼斯·克苏提克
项目地点：奥地利，维也纳
项目面积：100平方米
完成时间：2011年
摄影师：© 李·提兹

On the way from a nostalgic Parisian bistro to a cool, New York-style deli, a mixture of strongly alienated classicistic details and pop art-associated elements dominates the concept. The broad product range has been atmospherically translated into the room design. A colourful, placative one of a kind wallpaper with maniristic drawings embraces the room, forming a cheerful, ironic basis of design. In further steps, which have been applied as a branding in all corporate identity elements that have no connection to the room, the pattern becomes an essential element of the brand Nascha's.

Thanks to its distinctiveness, the architecture becomes a carrier of brand development. In a precise planning process, all product carriers have been reduced, in a functional manner, to the essential, while at the same time having been formed, in terms of design, in a very detailed manner. By doing so, a voluntary interpretation of the classic and well-known forms has been given priority. Colours have been reduced to black and white so as to draw attention to the gay world of colours of the products in an optimal way. A warm and not high-tech product illumination emphasises a familiar atmosphere. Added used vintage furniture and luminaires from the world of industry bestow upon the composition, which can be basically judged as elegant, a shabby chic touch, ensuring surprising contrasts.

This makes the boundary between New and Old fade, with a playful timelessness being created. The design developed in this manner has a nostalgic yet contemporary effect, being always on the borderline between irony and seriousness.

项目将一家怀旧的巴黎小酒馆改造成为一个纽约风格的时尚食品店,其中混合了变异的经典细部元素和现代艺术元素。广泛的经营范围被纳入了空间设计之中。五颜六色而又独一无二的墙纸上绘满了有趣的图案,形成了愉悦的设计基础。未来,娜莎食品店将会把这些图案推广成为该品牌的基本元素。

建筑的独特性使其成为了品牌发展的载体。在精确的规划流程中,产品载体的功能性被减少,更注重设计、造型等元素。经典而著名的造型拥有优先设计权。室内环境的色彩被缩减为黑白两色,以将人们的注意力吸引到五颜六色的商品之上。温馨的非高科技照明突出了亲切的居家感。来自工业时代的古旧家具和灯具完善了设计,增添了优雅而复古的气息,形成了强烈的对比。

设计在新旧之间建立的界限,形成了回味无穷的经典永恒,怀旧而有不失现代效果,幽默而又严肃。

1. Interior view
2. Interior viewed from the outside
3-4. Interior display detail
1. 店铺内景
2. 从店外看店铺内部
3、4. 店内陈列设计细节

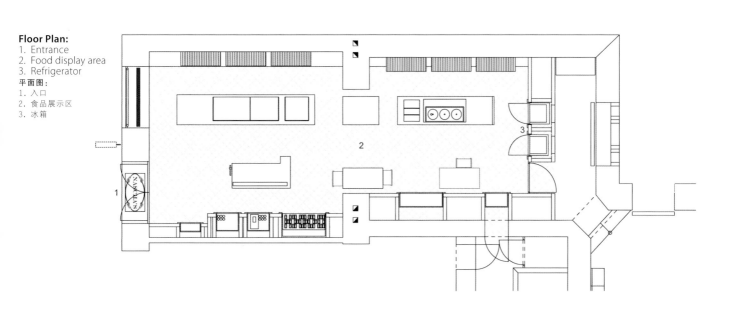

Floor Plan:
1. Entrance
2. Food display area
3. Refrigerator

平面图：
1. 入口
2. 食品展示区
3. 冰箱

5-7. Interior display detail
5~7. 店内陈列设计细节

Candy Room
糖果屋

Designer: RED DESIGN GROUP
Location: Melbourne, Australia
Project area: 50m²
Project year: 2010
Photographs: © Richard Kendall
设计师：红色设计集团
项目地点：澳大利亚，墨尔本
项目面积：50平方米
完成时间：2010年
摄影师：© 理查德·肯德尔

An exclusive, boutique candy store designed by the RED DESIGN GROUP — Australia's leading retail designers.

The founders of Candy Room approached RED DESIGN GROUP with a brief for a store that had to be edgy, humorous and uniquely charismatic. It was not to be simply a shop; it had to be a destination and an experience. Their new venture 'The Candy Room', located in the heart of Melbourne's CBD has a design that toys with the concept of illusion and draws the inner child out of the customer using a strong connection with childhood, fantasy, fiction and of course, sweets. Being strongly influenced by the idea of designing a playful, simple and somewhat illusional space for the Candy Room, the exaggeration of a 'room' idea was formulated. The application was to use line artwork on white space to represent a room.

Everything including the fixtures is painted in white, while graphically applied line artwork produce the suggestive elements of a room.

A kitchen splashback is drawn complete with a boiling pot on the stove or a framed portrait of one of the kids, tables are actually white boxes with black line drawings to appear as display fixtures.

RED was also responsible for the branding and all the packaging throughout the store. Allowing the space to be predominately white allowed the colours of the confectionery to dress the space. In a sense, the interior design for the Candy Room creates a fantasy and experience of a room without creating one.

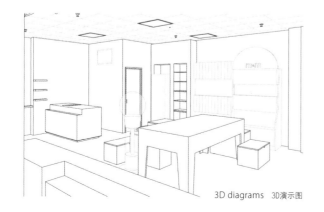

3D diagrams 3D演示图

1. Candy display shelf
2. Cash desk
3-6. Candy display design

1. 糖果展示架
2. 收银台
3-6. 糖果展示设计

第五章 食品展示_ 199

一间独特的,设计精美的糖果屋,由澳大利亚领先的零售设计公司红色设计集团所设计。

糖果屋的创始人带给红色设计集团的基本设计理念是,前卫、幽默和独特的魅力。它并不单纯是一个商店;而要成为一个目的和体验。他们新的合资企业"糖果屋"位于墨尔本中央商务区的核心地带,设计上稍稍运用概念上的错觉并且使用童年、幻想、故事,当然还有甜食来引出顾客们对童年的回忆。深受设计了一份俏皮、简单、有点幻想的空间的观点的影响,糖果屋的设计中,夸张的房间概念得以实现。通过在白色空间内使用线性图画来表现这一房间。

包括固定装置在内的一切被漆成了白色,而图形化的线性图画则给人带来房间的感觉。

厨房防溅板被画在了墙上,并且还画着一在炉子上沸腾的锅或一个孩子们的相框,由画着黑色线条的白色盒子构成的桌子充当了展示架。

红色设计集团还负责整个店铺的品牌和包装设计。让糖果的色彩来装饰这个纯白的空间。从某种意义上说,糖果屋的室内设计产生了一个充满幻想和体验的房间而非创建它。

_Chapter Five: Food Display

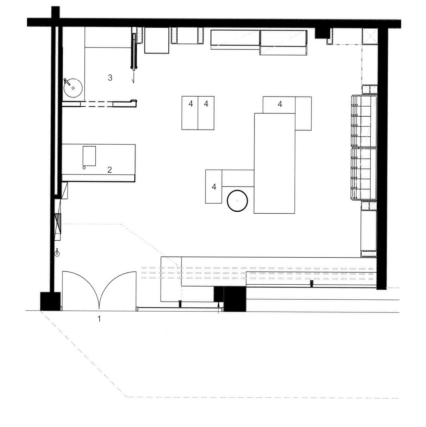

Floor Plan:
1. Entry
2. Cash counter
3. Staff
4. Display boxes

平面图：
1. 入口
2. 收银台
3. 员工区
4. 展示台

The Candy Stop Coyoacan
科约阿坎糖果店

Designer: ROW Studio – Álvaro Hernández Félix, Nadia Hernández Félix, Alfonso Maldonado Ochoa
Location: Mexico
Project area: 40m²
Project year: 2011
Photographs: © Sófocles Hernández for ROW Studio (Copyright ROW Studio)

设计师：ROW工作室——阿尔瓦罗·赫尔南德斯·菲利克斯、娜迪亚·赫尔南德斯·菲利克斯、阿尔凡索·马尔多纳多·奥克亚
项目地点：墨西哥
项目面积：40平方米
完成时间：2011年
摄影师：© 索福克勒斯·赫尔南德斯（图片版权：ROW工作室）

The Candy Stop Coyoacan is the first branch of a new candy store's brand in Mexico. Located in a traditional neighborhood in southern Mexico City, the design stands between the straight-forwardness and familiarity of the product and the desire of an eye catching proposal would attract customers; the middle ground is between the traditional vibe of the area and the contemporary feel of a new brand.

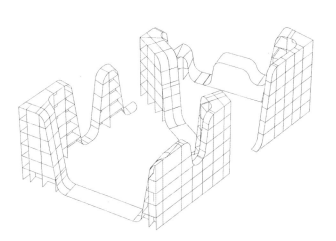

The entire exhibition is contained in an undulated rack that becomes a continuous purple ribbon that fuses with the floor and main desk in different moments. The same colour is used to frame the transitions between spaces and the main entry to the store.

An existing upper floor was partially demolished to create a double height area at the entrance enhancing a much needed sense of spatiality to the storefront. The upper surfaces of the walls are tilted to the inside in a quasi pyramidal shape, sharpening even more the perceivable height by elongating the visuals. This surfaces are completely covered by a repetitive pattern of abstracted gummy bears designed by industrial designer Ariel Rojo acting as a loose reference to the patterns and textures common in traditional architecture.

The materials remind of the nature of the product: solid, reflective and continuous surfaces that are both easy to clean and strikingly vibrant. An eye candy for sweets.

204 _Chapter Five: Food Display

1. Lighting design of the entrance area
2-4. Interior product display detail
1. 入口灯光设计
2~4. 店内商品陈列细节

科约阿坎糖果店是该糖果零售品牌在墨西哥的第一家分店，位于墨西哥城南部的一个传统社区之内。店铺设计介于直观表达和对产品的熟悉度之间，试图吸引顾客的眼球。设计融合了该区域的传统氛围和新品牌的现代感。

整个展示区设在一个波浪形的架子上，仿佛一条与地板和柜台融合在一起的连续紫色缎带。空间之间的过渡区与店铺主入口采用了同样的色彩。

原有的二楼被部分拆除，在入口营造出一个双层楼高的区域，提升了店面所需的空间感。墙壁的上部向内倾斜，类似金字塔形，在视觉上增强了高度感。墙面上完全被重复的抽象小熊软糖图案所覆盖，与传统建筑中的图案和纹理遥相呼应。

材料让人想起了产品的特质：封闭、反光而连续的表面便于清洁而充满活力，堪称视觉糖果。

Floor Plan:
1. Entrance
2. Display area

平面图：
1 入口
2 展示区

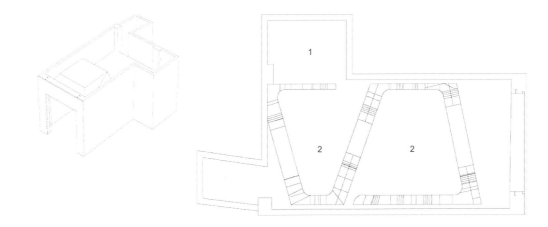

Chapter 6: Design Guidelines

第六章：设计规范

Walls and Ceilings
墙壁和天花

208 Wall Construction
墙壁

208 Wall Finishes for Food Preparation Areas
食品准备区墙面装饰

210 Ceilings
天花

210 Pipes, Conduits and Wiring
管线布置

Floors
地面

211 Floor Finishes for Food Preparation Areas
食品准备区地面饰面

211 Food Storage Areas
食品存放区地面装饰

212 Coving
脚线

212 Plinths
基座

Ventilation
通风

212 Natural Ventilation
自然通风

213 Mechanical Ventilation
机械通风

213 Filters
过滤装置

213 Food Shop Exhaust Hood Airflow
排烟气流

214 Storage Racks
存储架

214 Discharge Point
排气点

214 Dining Areas
就餐区

Sewage and Wastewater Disposal
废水及污水处理

Storage Facilities
存储空间

214 Dry Goods and Food Packaging Materials
干货及食品包装材料

215 Cleaning Chemicals and Equipment
清洁化学品和设备

215 Clothing and Personal Effects
员工服饰存放

215 Office Materials
办公器材

Above from left: Wall detail of a food shop
上图,从左至右:食品店墙面细节

WALLS AND CEILINGS

Wall Construction
Walls are to be solid and of framed or preformed panel construction where voids can be filled with a suitable material. Joints between preformed panels must be filled and finished flush with the surface of the sheeting material. Cover strips are not permitted in food preparation areas as they allow dirt and grease to accumulate.

Wall Finishes for Food Preparation Areas
Walls are to be finished with a light coloured, high gloss, washable and easy-to-clean surface. Walls in food preparation areas are to be finished with an approved material such as:
- glazed tiles (not suitable for wash down areas) – preferably laid to a minimum height of 2 metres;
- stainless steel or aluminium sheeting;
- acrylic or laminated plastic sheeting;
- polyvinyl sheeting with welded seams;
- pre-formed panels, villa board or compressed AC sheeting;
- trowelled cement (polished) may be appropriate in some circumstances.

墙壁和天花

墙壁
框架墙板结构内部需要采用适合的材质填充,确保其稳定与坚固性。墙板之间的接缝必须在填充饰面之后与整体结构保持水平。食品准备区域内的墙壁不可以采用覆盖条形式的结构饰面,避免灰尘和油污沉积。

食品准备区墙面装饰
墙面应当采用淡色系、高光材质来装饰,保证其易清洁性。
可用材质包括:
— 釉面砖(不适用于清洗区),墙壁的高度应当不低于2米
— 不锈钢或铝板
— 亚克力板或层压塑料板
— 带有焊接缝的聚氯乙烯板
— 饰面板、维纳板、压缩板
— 水磨石(在一些特殊范围内使用)

Suitability of Wall Finishes 墙面适合材料

Finish 装修	Wet Washed Areas 清洗区	Food Preparation 食品准备区	Vegetable Preparation 蔬菜准备区	Servery 上菜处	Store Room 存储区	Chillers/Freezers 冷却/冷冻区	Bin Store 箱子存储	Dining Areas 就餐区
Stainless steel 不锈钢	Yes 是	Yes 是	Yes 是	Yes 是	Yes 是	Yes 是	Yes 是	Yes 是
Ceramic tiles 瓷砖	Yes 是	Yes 是	Yes 是	Yes 是	Yes 是	Yes 是	Yes 是	Yes 是
Vinyl sheets 乙烯板	Yes 是	Yes 是	Yes 是	Yes 是	Yes 是	Yes 是	Yes 是	Yes 是
Painted plaster 彩绘石膏					Yes 是		Yes 是	Yes 是
Feature brick 功能砖								Yes 是
Aluminium sheet 铝板	Yes 是	Yes 是	Yes 是	Yes 是	Yes 是	Yes 是	Yes 是	Yes 是
Steel sheet 钢板						Yes 是		
Trowelled cement 抹平的水泥		Yes 是	Yes 是	Yes 是	Yes 是	Yes 是	Yes 是	Yes 是
Wood panelling 木镶板								Yes 是
Painted brickwork 彩绘砖					Yes 是		Yes 是	Yes 是
Concrete 混凝土					Yes 是		Yes 是	Yes 是
Pre-formed panels 预制板	Yes 是	Yes 是	Yes 是	Yes 是	Yes 是	Yes 是	Yes 是	Yes 是

Any finish continued above ceramic tiles must be finished flush with the tiles to prevent from dirt. Architraves, skirting boards, picture rails or similar protrusions on the walls in food preparation areas are not recommended.

Walls at the rear of cooking appliances must be surfaced with an impervious material, such as stainless steel, which extends from the canopy to the floor. Where a cooking appliance is sealed to the wall, the material must be lapped over the top edge of the appliance to provide a grease and vermin proof seal. Cooking appliances must only be sealed to walls made of a non-combustible material.

Splashback walls at the rear of benches, sinks and hand basins must be surfaced with an impervious waterproof material to a minimum height of 300mm. In wet areas, the bottom plate in all timber framed partitions in food preparation areas must be placed on a "dwarf" wall constructed of concrete or similar material, and constructed approximately 70mm above the floor.

瓷砖主体结构无论采用任何材质饰面都必须保持表面水平，以防止灰尘和油污沉积。食品准备区墙面上不应出现楣梁、衬板、挂镜线等凸起结构。

烹饪设备后方的墙面（从棚顶到地面位置）必须使用防水材质来整饰，比如不锈钢这类材质。烹饪设备如果固定在墙壁上，则墙壁饰面的部分必须高出设备本身的高度，以防止油污沉积导致的环境污染。除此之外，烹饪器具只能被密封在墙壁内，且墙壁必须采用非可燃材质打造。

操作台、清洗池和洗手盆后面的防溅墙必须采用防水材质饰面，高度不得低于3米。在食品准备区内，木框架墙板底板必须建在水泥（或者其他性质相同的材质）矮墙之上，高出地面约70毫米。

Suitability of Ceiling Finishes 天花板适合材料

Finish 装修	Wet Areas 清洗区	Vegetable Preparation 蔬菜准备区	Servery 上菜处	Store Room 存储区	Chillers/ Freezers 冷却/冷冻区	Bin Store 箱子存储	Dining Areas 就餐区
Painted plaster 彩绘石膏	Yes 是	Yes 是	Yes 是	Yes 是		Yes 是	Yes 是
Steel sheet 钢板	Yes 是	Yes 是	Yes 是	Yes 是		Yes 是	Yes 是
Trowelled cement 抹平的水泥	Yes 是	Yes 是	Yes 是	Yes 是		Yes 是	Yes 是
Wood panelling 木镶板							Yes 是
Concrete 混凝土	Yes 是	Yes 是	Yes 是	Yes 是		Yes 是	Yes 是
Pre-formed panels 预制板	Yes 是	Yes 是	Yes 是	Yes 是		Yes 是	Yes 是
Acoustic panels 隔音板							Yes 是
Decorative panels 装饰板							Yes 是

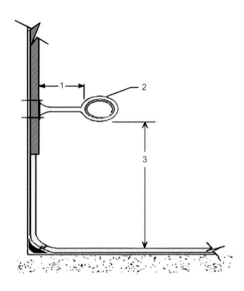

Pipes and Conduits Methods:
1. Minimum 25mm
2. Metal bracket
3. Minimum 100mm

管线设计：
1. 最小25毫米
2. 金属支架
3. 最小100毫米

Ceilings

The ceiling height in a food shop must not be less than 2.4m. Ceilings must be free of open joints, cracks and crevices. The intersection of walls and ceilings are to be tight jointed, sealed and dust-proof. The ceiling must be finished with a washable and impervious material. Ceilings must be finished in a light color to facilitate cleaning. Approved materials include:
- fibrous plaster, plasterboard;
- fibrous cement;

Drop in panels are not to be used in food preparation or display areas.

Pipes, Conduits and Wiring

Pipes, conduits and wiring must be concealed in floors, plinths, walls and ceilings, or fixed on brackets providing at least 25mm clearance between the pipe and adjacent surfaces. Service pipes, conduits and wiring are not to be placed in the recessed toe space of plinths or equipment.

天花

食品店内的天花板高度不能低于2.4米，必须保证无开口或裂缝，与墙壁之间的接缝需要坚固黏合并且确保防尘。天花板须采用淡色的、易清洗并且防水材质饰面。
可选用材质如下：
—纤维石膏板、石膏板
—纤维水泥板
此外，食品准备和展示区不可使用托板结构。

管线布置

管线必须隐藏在地面、基座、墙壁和天花下。如若固定在支架上，确保其与周围表面保持至少25毫米的空隙。特别注意的是，管线一定不能放置在底座或设备处。

Suitability Of Floor Finishes 地面适合材料

Finish 装修	Water Areas 清洗区	Food Preparation 食品准备区	Vegetable Preparation 蔬菜准备区	Servery 上菜处	Store Room 存储区	Chillers/ Freezers 冷却/冷冻区	Bin Store 箱子存储	Dining Areas 就餐区
Stainless steel non-slip 防滑不锈钢	Yes 是	Yes 是	Yes 是	Yes 是	Yes 是	Yes 是	Yes 是	Yes 是
Ceramic tiles 瓷砖	Yes 是	Yes 是	Yes 是	Yes 是	Yes 是	Yes 是	Yes 是	Yes 是
Quarry tiles 方砖	Yes 是	Yes 是	Yes 是	Yes 是	Yes 是	Yes 是	Yes 是	Yes 是
Steel trowel case hardened concrete 钢筋混凝土			Yes 是		Yes 是	Yes 是	Yes 是	Yes 是
Carpet/ carpet tiles 地毯/地毯砖								Yes 是
Wooden flooring 木地板								Yes 是
Poly vinyl sheet 聚氯乙烯板	Yes 是	Yes 是	Yes 是	Yes 是	Yes 是	Yes 是	Yes 是	Yes 是
Vinyl tiles 乙烯塑料地板			Yes 是	Yes 是	Yes 是	Yes 是	Yes 是	Yes 是
Plastic matting 塑胶地板				Yes 是				Yes 是
Cork tiles 软木砖								Yes 是
Epoxy resin 环氧树脂	Yes 是	Yes 是	Yes 是	Yes 是	Yes 是	Yes 是	Yes 是	Yes 是

FLOORS

Floor Finishes for Food Preparation Areas

Floors must be finished with an approved material and laid to a smooth surface, free from cracks and crevices. Floor finishes should be light colored to facilitate effective cleaning.

Floors are to be finished with one or a combination of the following materials:
-sealed quarry or ceramic tiles;
-stainless steel, non-slip;
-laminated thermosetting plastic sheeting;
-epoxy resin;
-steel trowel case-hardened concrete or similar impervious material;
-floor tiles with epoxy grout and finished flush with the surface of the tiles.

Food Storage Areas

Floors in storage areas for non-packaged food must meet the same requirements as floors in food preparation areas. Floors in storage areas for packaged food must have an impervious finish.

地面

食品准备区地面饰面
地面必须采用特定的材质装饰，确保其表面光滑，没有裂缝。地面装饰材质应当确保使用淡色系而且易于清洗。

可使用的材料如下：
— 石材或瓷砖
— 防滑不锈钢
— 热固性塑胶板
— 环氧树脂
— 硬化混凝土、防水材质
— 环氧灌浆地砖（保持表面平整）

食品存放区地面装饰
无包装食品存放区的地面装饰要求必须与食品准备区的地面相同；已包装食品存放区的地面必须确保具有防水性能。

CHAPTER SIX DESIGN GUIDELINES 第六章 设计规范

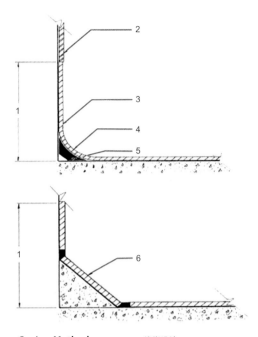

Coving Methods:
1. Minimum 75mm
2. Wall sheeting or tiles
3. Commercial grade vinyl
4. Minimum 25mm radius
5. Pre-formed backing piece
6. Tile- minimum 50mm in width and splayed 45°

线脚设计：
1. 最小75毫米
2. 护墙板或瓷砖
3. 商业级乙烯
4. 半径最小25毫米
5. 预制包装件
6. 瓷砖最小厚度50毫米，成45度角

Plinth Methods:
1. Splashback height min 300mm
2. Benchtop sealed to equipment
3. Equipment sealed to wall and plinth
4. Same as floor finish
5. Sealed
6. Plinth height min 100mm

基座设计：
1. 溅射高度最低300毫米
2. 密封在设备中的台面
3. 密封在墙和基座中的设备
4. 与地面装修相同
5. 密封的
6. 基座高度最低100毫米

Coving

Coving should be provided in new premises in areas where floors are intended to be or likely to be cleaned by flushing with water. Coving may also be required in existing premises, especially where cleanliness is an issue.

Plinths

Plinths can be used to hold heavy equipment that is unable or difficult to move for cleaning. Plinths are to be constructed to meet the same specifications as follows: must be solid; without voids and be an integral part of the floor; have the same top area finish as the floor; be rounded at all exposed edges and coved; be approximately 100 mm high. The base of the equipment is to be sealed to the plinth and overhung to prevent liquid, food or floor washing to access underneath the equipment. Alternatives to the use of plinths include metal legs, castors and brackets.

VENTILATION

Food shops must have sufficient natural or mechanical ventilation to effectively remove fumes, smoke, steam and vapours from the food shops.

Natural Ventilation

Natural ventilation is only suitable where there is little or no cooking that generates steam or greasy air. The shops must have openings, such as doors, windows and/or vents open to a clean environment.

脚线
在新建的食品店内，墙面和地面、天花板的接合处应当设计出脚线结构，便于用水清洗。这一结构对已经存在的食品店同样适用。

基座
基座用于放置那些难于移动或者不宜清洗的设备，它的构造要求需要满足如下的条件：坚固耐用、无中空，并且应当与地面能够融为一体、与地面采用相同的材质进行布置、所有的边缘要求呈现出圆弧形，并且高度保持在约为100毫米为最佳。除此之外，设备的底部要求与基座黏合，用来防止液体等物质侵蚀设备。基座的类型包括金属腿、脚轮以及支架。

通风
食品店内必须确保空气流通（自然或机械），以有效排除烟尘、蒸汽等。

自然通风
自然通风只适用于产生较少的油烟和蒸汽的空间。食品店内必须安装足够的门、窗等结构，便于空气流通。

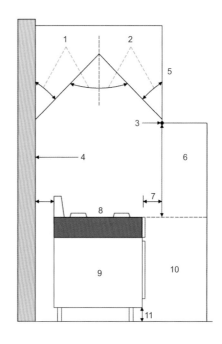

Mechanical Ventilation System:
1. No greater than 30° from vertical
2. No greater than 40° from vertical
3. Drain plug
4. Metal wall sheeting
5. Exhaust hood
6. Maximum 1200mm
7. Min 150mm
8. Heat source
9. Cooking appliance
10. Minium 2000mm
11. Minium 150mm

机械通风系统：
1. 不超过垂直30度角
2. 不超过垂直40度角
3. 排水塞
4. 金属护墙板
5. 排气罩
6. 最大1200毫米
7. 最小150毫米
8. 热源
9. 烹饪设备
10. 最小2000毫米
11. 最小150毫米

Mechanical Ventilation

A commercial kitchen must have a mechanical ventilation system that complies with local standards where any single apparatus has:
- a total maximum electrical power input exceeding 8 kilowatts (kW), or
- a total gas power input exceeding 29 megajoules per hour (MJ/h), or the total maximum power input to more than one apparatus exceeds:
- 0.5kW electrical power for each 1m² of floor area of the room or enclosure, or
- 1.8MJ gas for each 1m² of floor area of the room or enclosure dishwashers and other washing and sanitizing equipment that vent steam into the area to the extent that there is, or is likely to be, condensation collecting on walls and ceilings equipment installed on the shop after the mechanical ventilation system has been designed and installed must not impair the efficiency of the system or the natural ventilation

Filters

Canopies are to be fitted with grease filters which can be cleaned easier.

Food Shop Exhaust Hood Airflow

The airflow required for a food shop exhaust hood will depend on the:
-hood type;
-cooking process;
-length of the hood;
-inside perimeter of the hood over all exposed sides;
-height of the hood above cooking appliances.

机械通风

商业厨房必须配备机械通风系统，遵循当地相关标准，具体要求如下：

独立通风设备：
—电力输入最大值应超过8000瓦；
—气体输入功率应超过29兆焦耳/小时；

多个通风设备：
电力输入每平方米空间500瓦；
气体输入每平方米空间1.8兆焦耳；
洗碗机或其他清洁设备产生的蒸气需通过墙壁或天花上安装的专门设备排出，但一定要确保其不能妨碍机械通风设备或自然通风。

过滤装置

顶篷内应安装油烟过滤设备，便于清洁。

排烟气流

食品店内排烟气流设计需考虑以下方面：
— 排烟类型
— 烹饪过程
— 排烟管道长度
— 排烟设备内部规格
— 排烟设备距烹饪设备高度

CHAPTER SIX DESIGN GUIDELINES 第六章 设计规范

Storage Racks
Storage racks are not to be fitted above cooking and heating equipment as they can obstruct the airflow and trap droplets of oil.

Discharge Point
Effluent discharge is to be vertical at a minimum velocity of 5 metres per second (m/s) at the discharge point. The point of discharge is to be:
- 1m above the ridge of a pitch roof;
- 3m above a flat roof;
- 6m from a property boundary;
- 6m from any air intake, natural ventilation or opening.

Exhaust systems with a flow rate not exceeding 1,000 litres per second (L/s) may receive a relaxation on the location of the discharge point. No exhaust can discharge over adjoining properties or where the discharge is less than 3m above any pedestrian thoroughfare including an accessible roof area. Exhaust ventilation for wood-fired and solid fuel cooking equipment needs to be separate to other ventilation systems and must not be combined with systems serving grease appliances, or oil generating or oil-heat appliances.

Dining Areas
Dining areas must be ventilated by natural or mechanical methods.

SEWAGE AND WASTEWATER DISPOSAL
The food shop must have a sewage and wastewater disposal system that effectively disposes of sewage and wastewater, and is constructed and located so that it does not contaminate food or the water supply.

Food shops must provide appropriate facilities and plumbing infrastructure to ensure that sewage and all wastewater generated are disposed appropriately. The food shops must be designed in accordance with the following criteria:
- The design (hydraulics plans) and installation of sanitary plumbing and drainage must comply with the local standards and be approved by your local water authority;
- Installation and maintenance of a grease trap designed to filter

存储架
存储架不应安装在烹饪及加热设备上方，避免阻挡气流和油滴。

排气点
废气排放点应呈垂直方向设置，应达到最低5米/秒的速率，其设置应满足如下要求：
— 位于斜屋顶脊部1米之上
— 平屋顶3米之上
— 距屋顶边缘6米
— 距空气补充设备和自然通风口6米

排气系统流速如不超过1000升/秒，对于排气点的设置要求会相对宽松。任何废气不可通过相邻建筑排放，排气点与人行广场的距离不少于3米。木材或固体燃料产生的废气排放应与其他通风口分离开来。

就餐区
就餐区应当采用自然或者机械通风的方式，使空气流通。

废水及污水处理
食品店内必须具有污水处理系统，用于有效排放废水和污水，要求其不能污染食品和防碍净水供应。

食品店内须安装合适的排水管道系统，确保产生的废水和污水适当排放，具体要求如下：
— 卫生管道和排水系统的设计和安装必须满足当地水务部门的相关要求
— 要求安装用于过滤烹饪过程中产生的油脂的油脂收集设备
— 排水系统和油脂收集设备的入口应设在远离食品的区域内，同时确保其密封严实，阻挡害虫进入
— 为确保有效的地面清洁，冷藏室应配备室外地面污水排放管道
— 液体废料收集设备必须与排放管道相连

存储空间
干货及食品包装材料

grease and oil generated from the food business operations (where applicable) is required.
- Wastewater generated from mop buckets, cleaning mops and other cleaning activities must be disposed of in a cleaner's sink or other approved facility;
- Access openings to the sanitary drainage system and grease traps must not be located in areas where there is a risk of food contamination. Note all access points to grease arrestors are to have a tight fitting lid that will not buckle, warp or rust to prevent the entry of pests and vermin;
- To allow effective floor cleaning procedures, cold rooms should have an external floor waste drain located adjacent to the door;
- Equipment generating liquid waste must be connected to an approved tundish for correct discharge,.

STORAGE FACILITIES
Dry Goods and Food Packaging Materials
Adequate storage must be provided for dry goods and packaging materials in a sealed and lined, vermin-proof room with approved flooring.

Cleaning Chemicals and Equipment
Chemicals, cleaning equipment, pest control chemicals and equipment are to be: enclosed in cupboards located away from the preparation and storage of food where there is no likelihood of stored items contaminating food or food contact surfaces; designated for that use only.

Clothing and Personal Effects
Adequate facilities must be provided for staff to store personal belongings that consist of either a change room or enclosed cupboards for the storage of clothing and personal belongings, located away from the food preparation and storage areas.

Office Materials
Storage of paper work and other materials associated with the administration of the business must be stored in a room designated for that use or in enclosed cupboards or drawers designated for that use, separate from food preparation and storage areas.

干货及包装材料必须配备足够的存储空间，同时确保空间内整齐并防病虫侵害。

清洁化学品和设备
清洁化学品、防蚊虫化学品以及设备存放应满足如下要求：放置在密闭的橱柜内，并远离食品准备区和存放区，确保不能对食品造成危害；放置在专门用途的空间内。

员工服饰存放
食品店内必须提供足够的空间供员工存放个人物品，可以是专门的更衣室，也可以是单独的橱柜，要求远离食品准备区和存放区。

办公器材
与食品店经营相关的材料和文件必须存放在专门的空间内，或封闭的橱柜和抽屉内，要求远离食品准备区和存放区。

Chapter 7: Fixtures, Fittings & Equipements
第七章：家具、装置及设备

Equipment for Food Preparation and Storage
食品准备与存放设备

218 Temperature Gauges
温度计

218 Chilled and Frozen Storage
冷冻设备

218 Cold and Frozen Storage Rooms
冷藏和冷冻储存室

219 Preparation, Cooking, and Hot and Cold Display
食品准备、烹饪及展示设备

219 Benches, Tables and Preparation Counters
桌子、椅子和备餐台面

220 Cooking Equipment
烹饪设备

220 Display Cabinets
展示橱窗

221 Supports for Equipment
设备支架

221 Windows and Ledges
橱窗和壁架

221 Miscellaneous
其他

Equipment for Cleaning and Sanitizing Facilities
清洁和消毒设备

221 Hot Water Supply
热水供应

222 Double and Triple Compartment Sinks
双格和三格水池

222 Double Bowl Sinks
双格洗涤盆

222 Food Preparation Sinks
食材清洗池

217

CHAPTER SEVEN FIXTURES, FITTINGS AND EQUIPMENTS 第七章 家具、装置及设备

EQUIPMENT FOR FOOD PREPARATION AND STORAGE

Temperature Gauges
Hand held probe thermometers that are accurate to +/- 1°C are to be used to measure the internal temperature of the food.

Chilled and Frozen Storage
All cold storage and cold display equipment must be large enough for the business to adequately store cold food and must keep potentially hazardous food at a temperature of 5°C or less. Refrigerators, cold rooms and blast chillers must be capable of reducing the temperature of potentially hazardous food in accordance with local standards. Freezers are to keep food frozen. The recommended temperature for frozen food is at least -15°C.

Cold and Frozen Storage Rooms
Cold and frozen food storage room walls are to be lined with a smooth and impervious material and all joints sealed. Floors are to be a smooth and impervious material, and coved at the floor to wall junction. Floors are to be graded to the door opening and to a floor waste located outside the room, which is connected to the sewerage network or effluent disposal system. Doors must be able to be opened from the inside and an alarm fitted in accordance with the local standards. Shelving is to be made of galvanized piping (with

食品准备与存放设备

温度计
手持式温度计用于测量食品的温度，需精确到正负1摄氏度。

冷冻设备
店内所有的食品冷冻和冷藏展示设备必须确保拥有足够大的放置空间，以满足存储量的基本需要。同时，确保潜在危害食品的温度保持在5摄氏度或者更低温度。冰箱、冷藏室以及急速冷冻柜必须能够达到满足当地相关的食品卫生等标准的需要。冷藏库用于保存冷冻食品，建议温度保持在零下15摄氏度或以下。

冷藏和冷冻储存室
冷藏及冷冻存储室内，墙壁和地面均需要采用光滑的和防水材料来建造，并且要达到接缝处坚固、黏合的基本要求。同时，地面到门口处应该呈现出一定的坡度；室外的垃圾处理管道也需要与污水排放系统顺畅地连通。店门必须能够向内开放，并且要安装警报设备以确保能够应对安全与临时疏散情况的发生等。商品的货架构造应当采用不锈钢或者其他符合标准的材质打造，确保墙面能够易于清洁和

Above: Sample equipments of a food shop
上图：食品店常用设备

sealed ends), stainless steel or other suitable materials; must be easy to remove for cleaning; is clear of walls for cleaning and maintenance; and the lowest shelf must be at least 250mm off the floor to allow for easy cleaning.

Preparation, Cooking, and Hot and Cold Display

All equipment for preparation, cooking and display must be constructed to be easily and effectively cleaned with no open cracks, crevices and joints where food and liquids can collect.

Hot and cold food storage and display units must be capable of maintaining food under temperature control (5°C or below, 60°C or above).

Benches, Tables and Preparation Counters

Benches and tables are to be constructed so that they are able to be easily and effectively cleaned and sanitized. Examples include laminated timber, plastic or stainless steel with pest-proof joints.

Benches fixed against a wall must be sealed to the wall with an appropriate material. Benches subjected to heat should be lined with stainless steel. Sandwich counters, used to prepare food in front of customers, must be fitted with a protective barrier between the customer and the food.

获得及时维护。除此之外，货架的底部与地面相隔至少需要250毫米的空间，避免藏污纳垢，更便于清洗地面。

食品准备、烹饪及展示设备
所有用于准备、烹饪和展示食品的设备均必须满足易于清洁的要求，并且确保无开口、裂缝，防止液体渗入导致的污染。

保温及冷藏存储和展示设备必须满足将食品的温度维持在60摄氏度以上和5摄氏度以下。

桌子、椅子和备餐台面
桌椅、台面等店内家具要满足易于清洁和维护的要求，比如：层压木材、塑料及不锈钢材质可以大量地运用。

固定在墙壁上的座椅必须要确保采用的是适当的材质。耐高温的座椅可以采用不锈钢的材质饰面。

备餐台面必须在顾客和食品之间安装一层隔断，与顾客距离不足1.5米的烹饪设备必须安装玻璃喷嚏挡，防止污染。

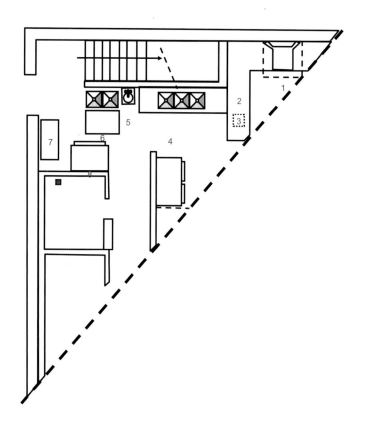

Sample Equipment Layout:
1. Warewashing machine
2. Stainless steel dish counter
3. Floor sink
4. 3 compartment sink
5. Hand sink
6. 2 compartment prep sink
7. Ice maker
8. Freezer

标准设备布局：
1. 洗碗机
2. 不锈钢碗柜
3. 地漏
4. 三格水槽
5. 洗手槽
6. 二格准备水槽
7. 制冰机
8. 冰箱

Preparation benches and cooking equipment less than 1.5m from customers must be fitted with sneeze guards constructed of glass or perspex, and designed to prevent contamination from customers. Equipment placed on bench tops must be:
- easy to move by one person
- raised above the bench top to allow easy access for cleaning
- sealed to the bench top

Cooking Equipment

Stoves and cooking appliances are to be kept clear of walls to enable access for cleaning or built into walls and completely pest proofed. Appliances must be either:
- placed apart to prevent grease and food accumulation;
- placed together with the gap between the appliances flashed or sealed to prevent food, liquid or grease accumulating, or placed on castors to allow the appliance to be moved for cleaning;

Appliances must be kept clear of cupboards or benches not used in connection with the cooking operation. Deep frying equipment must be thermostatically controlled to prevent a fire hazard from the overheating of cooking oils. Where cooking is carried out, the premises must be equipped with a suitable fire extinguisher or fire suppression system located near the cooking equipment as required by the relevant legislation.

台面上设备必须满足如下要求：
— 便于移动
— 与台面保持距离，便于清洁
— 与台面牢固结合

烹饪设备

烹饪设备放置必须要与墙壁保持一定距离，这样才能够确保卫生，或者将这些设备直接嵌入到墙壁内，防止昆虫等的侵蚀。
其他细节要求如下：
— 设备之间要求分隔开来，防止油污沉淀
— 设备放置在一起，之间的缝隙确保黏合、防止油污沉淀、设备放置在脚轮上，便于移动清洁

所有的设备必须和与烹饪整个过程无关的橱柜等设施保持一定距离。深度煎炸的设备必须具备温度控制设施，以免引起火灾。根据相关的需求，烹饪空间内必须配备消防设施。

展示橱窗

展示橱窗的滑动门的下部必须配备指示标识，但是要确保与底端的距离不要小于25毫米，以便能够保证卫生。

Sample Food Preparation Area Layout:
1. Floor/wall coving
2. Plinth not less than 100mm high
3. Impervious floor graded and drained
4. Fittings sealed to wall or 200mm clear of wall
5. Walls finished
6. Sealing between fittings
7. Legs 150mm minimum
8. No storage shelves below canopy
9. Splayed windowsill 300mm above preparation bench
10. Preparation bench – steel framed
11. Bottom shelf – min 250mm above floor
12. Mechanical exhaust ventilation canopy
13. Rigid smooth faced ceiling
14. Smooth trowelled cement
15. No timber door frames
16. Hand basin, hot and cold water mixing set
17. Soap and towel dispenser
18. Water and drainage pipes concealed in wall

标准食品准备区布局：
1. 地面/墙壁线脚
2. 基座不低于100毫米
3. 分流和排水的防水地面
4. 设备装进墙内或距墙200毫米
5. 墙面装修
6. 设备之间密封
7. 腿部——最小150毫米
8. 排风罩下方不可以有储物架
9. 操作台上方300毫米张开的窗台
10. 准备台——钢结构
11. 底层货架——地面以上最少250毫米
12. 机械排风罩
13. 光滑表面的天花板
14. 打磨光滑的水泥
15. 不可以使用木质门框
16. 冷热水混合的洗手盆
17. 香皂和毛巾
18. 装进墙壁的供水和排水管

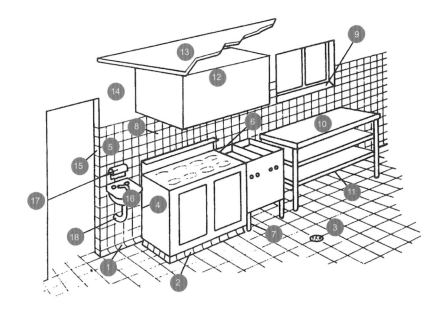

Display Cabinets

Sliding doors to display cabinets must have bottom guides or runners terminating not less than 25mm from each end of any door opening for easy cleaning.

Self-service food bars must be fitted with sneeze guards designed to prevent contamination (from a customer's mouth or nose) affecting the food. Window displays of wet foods, such as meat and fish, must be coved at all intersections and installed in accordance with the local standards.

Supports for Equipment

Including plinths, the following can be used to support heavy equipment: metal legs - are to be smooth and sealed to prevent the access of pests and be approximately 150mm high for easy cleaning; castors or wheels - must be capable of supporting and moving fully-loaded equipment; brackets – sinks, tubs, wash basins, tables, benches, shelving and similar fittings must be supported on stainless steel.

Windows and Ledges

Windows should be located at least 300mm above the bench, sink or hand basin. Ledges must be splayed at a 45° angle to prevent accumulation of dirt, food and grease.

自助食品展示台必须安装喷嚏挡，以防止污染。用于展示肉类和鱼类食品的橱窗的所有接缝处应当呈现弧线型，满足相关标准要求。

设备支架

以下结构可用于支撑较为沉重的设备：金属腿，要求表面光滑并且坚固、黏合，能够防止昆虫进入，最佳高度为150毫米，这样便于清洁；脚轮，要求能够支承并滑动较重的设备；支架，要求用于支承水池、桌椅、橱柜等，要求采用不锈钢材质打造。

橱窗和壁架

橱窗应位于椅子、水池、手盆等至少300毫米以上的上方；壁架应呈现45度角倾斜，防止灰尘、食品及油烟的沉淀。

其他

制冰机等设备应放在完全密封的空间内。食品传送设备表面要光滑并防水，无开口、裂缝、易清洁。

清洁和消毒设备

热水供应

所有清洁设备都必须保持持续的冷、热水供应。清

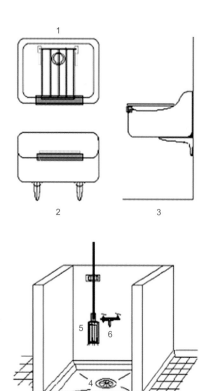

Cleaner's Sinks:
1. Top view
2. Front view
3. Side view
4. Floor drain to server
5. Mop
6. Hot and cold water

清洁水槽：
1. 顶部图
2. 正面图
3. 侧面图
4. 地漏
5. 拖把
6. 热水和冷水

Miscellaneous

Post-mix/syrup and ice machines must be located in a fully lined and sealed room (i.e. they must not be stored outside the food premises or in the open) constructed to the specifications of a food preparation/storage room. Food conveyors must be constructed of smooth impervious surfaces, free from cracks, crevices and open joints, with access provided for easy cleaning.

EQUIPMENT FOR CLEANING AND SANITIZING FACILITIES

Hot Water Supply

All equipment for cleaning and sanitizing is to be connected to a continuous supply of hot and cold potable water and to an approved drainage system. Sinks must be supplied with water at a temperature of not less than 54 degrees Celsius (°C) for washing and 77°C for sanitizing (if sanitizing takes place in the sink). The best temperature for washing utensils in the food service industry is between 54°C and 60°C. Temperatures higher than this tend to make food residue.

Double and Triple Compartment Sinks

Where a double or triple compartment sink is used for hot water sanitizing, rinsing baskets and heating elements capable of maintaining the water temperature at a minimum of 77°C are to

洗池要求热水温度至少为54摄氏度，消毒设备则应保持77摄氏度。对于食品店来说，餐具清洗最佳温度在54摄氏度和60摄氏度之间，过高的温度会造成烘焙食品残留。

双格和三格水池

用于热水消毒、清洗的双格或者三格水池应该保持热水温度的最低值为77摄氏度，其附近应需配备有滴水或烘干空间。在清洗的过程中，可以使用一些化学消毒剂。

双格洗涤盆

双格洗涤盆需满足如下要求：由不锈钢材质打造；最小规格为450毫米x300毫米x300毫米，便于能够洗涤较大的容器；两侧配备排水区；配备防喷溅挡板，与墙壁距离300毫米；洗涤盆上方的滴水架同样要求采用不锈钢材质打造。

食材清洗池

食品准备过程中，如食材需清洗，则必须配备清洗池，并确保与其他清洗池分离开来。与其他清洗池之间的距离应由相关人员确定，主要目的是避免食材污染。需考虑的因素包括食品制作流程、水池的规格等。

Sample Wash-up Area Layout:
1. Floor/wall coving
2. Casters to under-bench storage
3. Impervious floor graded and drained
4. Hot water heater sealed to wall
5. Walls finished
6. Shelving – 40mm clear of wall
7. Sink unit on metal frame
8. Thermometer
9. Garbage receptacle
10. Dishwasher with temperature indicating device
11. Legs - 250mm minimum
12. Bottom shelf - minimum 250mm above floor
13. Rigid smooth faced ceiling
14. Smooth trowelled cement
15. Water and drainage pipes concealed into walls
16. Hand basin, hot and cold water mixing set
17. Soap and towel dispenser
18. Mechanical exhaust ventilation canopy

标准清洗区布局：
1. 地面/墙壁线脚
2. 操作台下方存储柜的脚轮
3. 分流和排水的防水地面
4. 装入墙内的水加热器
5. 墙面装修
6. 货架——距墙面40毫米
7. 金属框架上的水槽
8. 温度计
9. 垃圾容器
10. 装有温度指示装置的洗碗机
11. 腿部——最小250毫米
12. 底层货架——地面以上最少250毫米
13. 光滑表面的天花板
14. 打磨光滑的水泥
15. 装进墙壁的供水和排水管
16. 冷热水混合的洗手盆
17. 香皂和毛巾
18. 机械排风罩

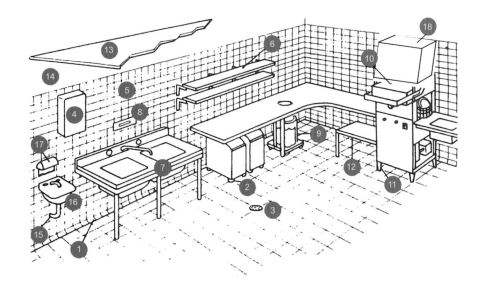

be provided. Loading space and draining or drying space is to be provided. Chemical sanitizing is permissible.

Double Bowl Sinks

Double bowl sinks must have the following requirements: be constructed of stainless steel; have a minimum bowl size of 450mm x 300mm x 300mm to enable cleaning of large pots and equipment; be fitted with a draining area at each end; have a splashback as part of the unit, 300mm off the wall; where draining racks are provided above sinks, they must be of stainless steel construction (preferable to have walls behind a drainage rack made of stainless steel sheeting or tiles to prevent damage to the wall).

Food Preparation Sinks

Where food preparation requires the washing of food and immersion in water, a designated food preparation sink must be provided for this purpose. Designated food preparation sinks must be separate from all other sinks. Separation distance between sinks is to be determined by authorized persons with regard to the implicated risk of food or food contact surface contamination. This may be considered in conjunction with the factors such as the operation flow of food production, the size and depth of the sinks concerned, the feasibility of any acceptable alternative engineering solutions, e.g. the provisions of a physical barrier between the sinks.

Index 索引

1+1=1 Claudio Silvestrin Giuliana Salmaso architects and planners
Unit 412 kings wharf
301 kingsland road, London e8 4ds
Tel: +44 7734 087430
E-mail: info@csgs.info

a l m project inc
5544 hollywood boulevard
Los Angeles, CA 90028
Tel: +1 323 570 0571
E-mail: studio@almproject.com

architektur denis kosutic
florianigasse 7/8 1080
Vienna, Austria
Tel: +43 699 19479990
E-mail: contact@deniskosutic.com

Baker Architecture + Design
E-mail: grijalva@bakerad.com

Carbon Design Associates Ltd.
Unit D 103 Lana House, 116 Commercial,
London E1 6NF, United Kingdom
E-mail: info@carbon-creative.com

Cinimod studio Ltd.
Unit 304, Westbourne Studios,
242 Acklam Road,
London W10 5JJ, England
studio +44 20 8969 3960
E-mail: enquiries@cinimodstudio.com

DOYLE COLLECTION Co.,ltd.
1-20-3-302, Ebisu-minami, Shibuya-ku,
Tokyo, Japan Zip.150-0022
Tel: +81-3-5734-1508
Fax: +81-3-5734-1509
E-mail: info@doylecollection.jp

GH+A
1100 avenue des Canadiens-de-Montréal
Suite 130, Montréal, Quebec
Canada H3B 2S2
Tel: +1.514.843.5812

GLAMOROUS Co., ltd.
2F, 2-7-25 Motoazabu, Minato-ku,
Tokyo 106-0046, Japan
Tel: +81 3-5475-1037
Fax:(+81)3-5475-1038
E-mail: info@glamorous.co.jp

Jonathan Clark Architects
3rd Floor, 34-35 Great Sutton Street
London EC1V 0DX
Tel: 020 7608 1111
E-mail: mail@jonathanclark.co.uk

KAMITOPEN Architecture-Design Office Co.,ltd.
TakefushiBld 2F, 2-4-6,
Asakusabashi, Taito-ku,Tokyo, Japan
Tel: +81(0)3-6240-9856
E-mail: yoshida@kamitopen.com

KARIM RASHID Inc.
357 West 17th St.
New York, NY 10011
Tel: 212.929.8657
Fax: 212.929.0247
E-mail: office@karimrashid.com

Minale Design Strategy
55, rue Danton
F-92300 Levallois Perret, Paris
Tel: +33 1 41 92 97 00
E-mail: info@minaledesignstrategy.com

minifie van schaik architects
top floor 181 swanston st
melbourne 3000, Australia
Tel: +613 9654 6326
E-mail: office@mvsarchitects.com.au

mooof design studio
28 Charoen-nakorn 10, Charoen-nakorn Road,
Klong San, Bangkok, Thailand 10600
Telephone/Facimile: +66(0)2 437 8977
mobile:+66(0)86 893 8758
E-mail: contactmooof@yahoo.com

MORRIS SELVATICO
Studio 303, 50 Holt Street
Surry Hills, Nsw 2010, Australia
Tel: +61 (0)2 9380 2380
Fax: +61 (0)2 8065 4018
E-mail: info@morrisselvatico.com

Nota Design International Pte Ltd.
422 Joo Chiat Road, Nota House, Singapore 427642
p: 63459182 f: 63446471
Paul Burnham Architect Pty Ltd.
Perth, Western Australia
Tel: 0439 095 865
E-mail: burnham@westnet.com.au

Pereira Miguel Arquitectos
Lisboa, Portugal
E-mail: info@pm-arq.com

Project Orange
1st Floor, Cosmopolitan House
10A Christina Street, London EC2A 4PA
Tel: +44 (0)20 7739 3035
Fax: +44 (0)20 7739 0103
E-mail: mail@projectorange.com

RED DESIGN GROUP
Level 12
160 Queen Street
Melbourne VIC 3000
Tel: 03 9693 2500
Fax: 03 9686 7110
E-mail: info@reddesigngroup.com.au

ROW//Studio
Palmas 1145
Lomas de Chapultepec
C.P 11000
Mexico, Distrito Federal
Tel: +52 (55) 47539565
E-mail: info@rowarch.com

Studio Alberto Re
via Abbadesse, 36
20124 Milan (Italy)
Tel: +39 02 6883876
Email: info@studioalbertore.com

Studio Linse
Keizersgracht 534
1017 EK Amsterdam
Tel: +31 (0)20 675 47 98
Fax: +31 (0)20 671 90 68
E-mail: info@studiolinse.com

VONSUNG
+44 (0) 207 650 8909
E-mail: info@vonsung.com

Zwei Interiors Architecture
Studio 3.06 / 87 Gladstone Street
South Melbourne, Victoria 3205
Tel: +61 3 9696 3104
Fax: +61 3 9696 3105
E-mail: info@zwei.com.au